Talking Green

This book is part of the Peter Lang Media and Communication list.
Every volume is peer reviewed and meets
the highest quality standards for content and production.

PETER LANG
New York • Washington, D.C./Baltimore • Bern
Frankfurt • Berlin • Brussels • Vienna • Oxford

Talking Green

Exploring Contemporary Issues in Environmental Communications

EDITED BY
Lee Ahern & Denise Sevick Bortree

PETER LANG
New York • Washington, D.C./Baltimore • Bern
Frankfurt • Berlin • Brussels • Vienna • Oxford

Library of Congress Cataloging-in-Publication Data

Talking green: exploring contemporary issues in environmental
communications / edited by Lee Ahern, Denise Sevick Bortree.
p. cm.
Includes bibliographical references and index.
1. Communication in the environmental sciences. 2. Environmental ethics.
I. Ahern, Lee. II. Bortree, Denise Sevick.
GE25.T35   333.701'4—dc23   2012026571
ISBN 978-1-4331-1791-6 (hardcover)
ISBN 978-1-4331-1790-9 (paperback)
ISBN 978-1-4539-0887-7 (e-book)

Bibliographic information published by **Die Deutsche Nationalbibliothek.**
Die Deutsche Nationalbibliothek lists this publication in the "Deutsche
Nationalbibliografie"; detailed bibliographic data is available
on the Internet at http://dnb.d-nb.de/.

The paper in this book meets the guidelines for permanence and durability
of the Committee on Production Guidelines for Book Longevity
of the Council of Library Resources.

Printed in the United States of America

This volume was made possible through the generous support of:

THE ARTHUR W. PAGE CENTER
FOR INTEGRITY IN PUBLIC COMMUNICATION

College of Communications
The Pennsylvania State University

TABLE OF CONTENTS

Talking Green: Exploring Contemporary Issues in Environmental Communications

Lee Ahern

Denise Sevick Bortree

In *Talking Green,* several leading environmental communications researchers—and one leading practitioner—analyze and explore some of the dominant obstacles confronting environmental advocates. The volume is designed to narrow the gap between research and application and provide insights to interested parties in both areas. There is no one "right answer" to the question of environmental communications strategies and tactics, just as there is no one "environmental communications." The wide range of interrelated issue areas that comprise environmentalism requires a wide range of approaches to audience research and campaign development and implementation. The goal here is to identify and explore some of the common questions and challenges that confront environmental communicators.

As relayed below, the book is informed and inspired in large part by the 2008 essay "Weathercocks & Signposts: The Environment Movement at a Crossroads" by WWF-UK Change Strategist Tom Crompton. As a springboard into a critical issue area, the Executive Summary of this essay is reprinted as Chapter 1 and serves as a point of departure for the chapters that follow. An Afterword by Dr. Crompton details the directions he has gone since "Weathercocks" was published and further contextualizes the work of our contributors.

Background and Purpose

Every day, thousands of people around the world engage in the study and practice of environmental communications. These passionate advocates understand that the principal barriers to the planet's greatest environmental challenges are not technological but social—the lack of public understanding and urgency about the issues, and, too often, the lack of collective political will to act. The world can take on its environmental problems with current technology, but it will never be successful without better communications campaigns.

Globally, the number of groups, associations and social organizations dedicated to environmental issues is huge and growing fast. Wiser Earth, an online community for the global environmental movement founded by the environmental advocate and author Paul Hawken, lists over 110,000 non-governmental organizations dedicated to social environmental justice and sustainability. This list does *not* include the multiple environmental public affairs officers at each level of government, for most of the governments in the world, and the thousands of environmental marketing, advertising and public relations professionals around the globe. In the area of education, thousands of courses have been developed partially or fully dedicated to advancing understanding of environmental communication, and each of the largest academic communications associations has a division or interest group focused on this growing and dynamic area.

These communicators and researchers are operating in a changing and challenging landscape. Environmentalism has evolved from a new social movement to an institutionalized collection of special interests, and the challenge has grown from local pollution to global warming. In addition, audiences are becoming more fragmented by changes in media delivery, and more politically polarized (in many countries) by the balkanization of news outlets. Information overload is engendering issue fatigue, and digital technology is altering what it means to "take action."

Although this is a vast overgeneralization of the recent developments within environmentalism, it is safe to say there is a growing consensus that despite a number of great success stories, new approaches are needed.

In 2004 Michael Shellenberger and Ted Nordhaus, two experienced environmental communications strategists, shook things up with their essay "The Death of Environmentalism: Global Warming Politics in a Post-Environmental World," arguing that environmentalism had to "die" so that it could be replaced by a new approach. Their vision moves the focus from environmental protection to sustainable economics and was more fully developed in their 2007 book *Break Through: From the Death of Environmentalism to the Politics of Possibilities* (see www.thebreakthrough.com for a great deal of useful additional information and).

Further reflecting the fundamental transformation of environmentalism is the division of the movement into "light green," "dark green" and "bright green" groups, a taxonomy developed by writer, activist and "Worldchanging" author Alex Steffen. In this construction light greens see environmentalism as a personal responsibility but are not dedicated to extreme public policy solutions, dark greens see the problem as socio-political and therefore support radical political change, and bright greens envision a transformation fueled by emerging energy and communications technologies (more at www.worldchanging.com).

In this atmosphere of change, environmental communicators were naturally looking for new ways to talk about new problems and challenges. Advances in the psychology of persuasion as applied to environmental appeals began to directly inform campaign strategy. In 2008 Dr. Tom Crompton published "Weathercocks," which reassessed current approaches to motivating environmentally-friendly behavior change, and challenged the analogy that "green" needs to be "sold" using a marketing approach. While critical of dominant environmental messaging strategies, it was also optimistic that more firmly grounded, values-based appeals can effectively produce the massive individual and social change required by global environmental challenges.

The editors combined the questions and challenges raised by "Weathercocks" with a national call for grant proposals issued by

the Arthur W. Page Center for Ethics in Public Communication at
Penn State University College of Communications. Each year, the
Page Center funds worthy research in the area of communications
ethics and practice, and in 2010 a special call was issued for studies that engaged with "Weathercocks," with the goal of building
knowledge about strategic environmental communication. The
grant-winning authors (along with other leading researchers working in the area) were then invited to submit their findings in the
form of a chapter for this volume. The results of this collaboration
comprise this book.

Chapter summaries

Following the reprint of the "Weathercocks" essay, Matthew C.
Nisbet, Ezra M. Markowitz and John E. Kotcher further explore
the problematic nature of several contemporary environmental
campaigns (Chapter 2: Winning the Conversation: Framing and
Moral Messaging in Environmental Campaigns). They offer an alternative approach reflective of emerging understanding of individual values and their impact on message reception and
interpretation. Recent research in the area of intuitive ethics
points to a range of moral foundations that underlie reasoning
about public policy and details how people with opposing socio-
political worldviews tend to emphasize or rely on different generalizable core beliefs. This contextualizes the impacts of values- and
identity-based environmental campaigns in the current highly po-
liticized social and media landscape in the US.

Charles T. Salmon and Laleah Fernandez took inspiration
from the "Weathercocks" essay to highlight the importance of considering unintended consequences of environmental campaigns
(Chapter 3: Biofuels and the Law of Unintended Consequences).
Through an empirical analysis of a set of influential articles on
biofuels in *Science* magazine, they examine how framing of consequences turned the tide of public opinion relative to an entire mul-
ti-billion-dollar industry. Many environmental public policy
debates are rife with potential unintended consequences. This
chapter illustrates how important it is to understand and appreci-

ate these potentialities; legitimation of unintended consequences can move public opinion, impact public policy in ways that can be difficult if not impossible to reverse.

The rhetorical dimensions of "greenwashing" are further explicated by Brant Short (Chapter 4: Greenwashing to Green Advocacy: The Environmental Imperative in Organizational Rhetoric). Short's analysis details how oft-criticized organizational rhetoric can represent a powerful positive voice for sustainability. He cites the environmental efforts of religious organizations as a case in point, providing further support for the idea that community-level opinion leaders are critical in engaging the significant attitude and behavior change required to combat the problem of global warming (a theme consistent with the argument of Nisbet et al. in Chapter 2).

Authors Michael Palenchar and Bernardo H. Motta (Chapter 5: Environmental Risk Communication: Right to Know as a Core Value for Behavioral Change) respond to Crompton's essay by arguing that "right to know" is fundamentally linked to environmental behavioral change. Palenchar and Motta discuss the relationship between motivations of the environmental movement and the value of right to know in the context of risk communication. The chapter offers insights into the importance of risk equity and intrinsic motivations in the effective implementation of right to know policies, and it acknowledges the need for an educated public in order to achieve significant change.

In "Weathercocks," Crompton argues that advocacy organizations could drive environmental action more successfully through communication strategies that use a values-based approach rather than those modeled on strategies of marketers. Three chapters in this book take a closer look at the marketing vs. values-based approach toward environmental communication. First, Janas Sinclair and Barbara Miller (Chapter 6: Public Response Before and After a Crisis: Appeals to Values and Outcomes for Environmental Attitudes) respond to Crompton's call for further exploration of the use of marketing approach with a chapter that defines marketplace advocacy activities of corporation and identifies the motivations and consequences of the campaigns. The chapter proposes a model for effective corporate environmental campaigns as a part of marketplace advocacy and offers suggestions for advocacy groups to minimize the impact of these campaigns. Sinclair and Miller argue that sponsor accountability, message trustworthiness, and audience motives play a key role in affecting audience attitude toward environmental issues.

Continuing the discussion of the marketing approach as a means to motivate environmental action, authors Harsha Gangadharbatla and Kim B. Sheehan (Chapter 7: Individual Factors and Green Message Reception: Framing, Lifestyles and Environmental Choices) argue that the marketing approach is not entirely without value in motivating behavior. Through an experiment with 329 participants, the authors find that those exposed to environmental messages have a higher intention toward environmental actions. Individual differences play a role as well, and the authors discuss the implication of these findings.

The chapter by Denise Bortree (Chapter 8: Pro-Environmental Behaviors through Social Media: An Analysis of Twitter Communication Strategies) takes a closer look at the marketing and values-based approaches by tracking the use of the strategies in social media communication by one of the world's largest environmental organizations, the Sierra Club. By examining the strategies used to promote public-sphere and private-sphere behaviors at the local, state, regional and national levels, she finds that both approaches are used, but the pattern of use could be altered to create a more effective communication strategy.

Finally, the focus on values, identity and moral foundations at the individual intuitive level and the emphasis on corporate social responsibility at the organizational level raise a set of ethical issues related to the use of modern advertising techniques in the pursuit of green consumption and pro-environmental behavior. Lee Ahern expands on this theme and develops an approach to ethical mapping designed to help communicators sort through the problematic nature of persuasive campaigning (Chapter 9: Evaluating the Ethicality of Green Advertising: Toward an Extended Analytical Framework). Indirect emotional appeals—the type of approach that may be the most effective in engaging intuitive moral foundations—can be ethically problematic. Inasmuch as emotional appeals persuade beyond conscious awareness, they can be seen as diminishing individual autonomy. By introducing the dimension of the public sphere (and the publicality of a specific issue-area), Ahern seeks to map out when these types of appeals are most problematic. The result is an extended framework that helps distinguish when emotional appeals are appropriate, and at the same helps evaluate when an environmental message can be considered ethically problematic greenwashing.

As the central metaphor in the "Weathercocks" essay suggests, we need to move beyond the observation of indicators of how environmental attitudes form and move on to erecting clear signposts for people to follow. As the following chapters reflect, a great deal about this critical issue area has been learned in recent decades through various and multiple social science disciplines. Translating these findings and advances into concrete communications campaigns takes not only skill but courage. As much as science has to offer, designing successful message strategies still involves a great deal of art. Advocates must choose which weathercocks should guide them in their placement of signposts. We believe the reflections and insights gathered here will stiffen the resolve of researchers and strategic planners to move forward with the next generation of environmental communications campaigning. The challenges, and the potential rewards, of this hard work are greater than ever.

CHAPTER ONE

Weathercocks and Signposts: The Environment Movement at a Crossroads

Tom Crompton

The following Executive Summary (originally published in 2008) is reprinted with permission of Tom Crompton and WWF-UK. Visit www.valuesandframes.org to view the full report along with other relevant material.

"It is no use saying, 'We are doing our best.' You have got to succeed in doing what is necessary."—Winston Churchill

As our understanding of the scale of environmental challenges deepens, so we are also forced to contemplate the inadequacy of the current responses to these challenges. By and large, these responses retreat from engaging the values that underpin our decisions as citizens, voters and consumers: mainstream approaches to tackling environmental threats do not question the dominance of today's individualistic and materialistic values.

"Weathercocks and Signposts" critically reassesses current approaches to motivating environmentally-friendly behavior change. Current behavior-change strategies are increasingly built upon an analogy with product marketing campaigns. They often take as given the "sovereignty" of consumer choice, and the perceived need to preserve current lifestyles intact. This report constructs a case for a radically different approach. It presents evidence that any adequate strategy for tackling environmental challenges will demand engagement with the values that underlie the decisions we make—and, indeed, with our sense of who we are.

The Marketing Approach to Creating Behavioral Change

Pro-environmental behavioral change strategies often stress the importance of small and painless steps—frequently in the expectation that, once they have embarked upon these steps, people will become motivated to engage in more significant behavioral changes. Often, these strategies place particular emphasis on the opportunities offered by "green consumption"—either using marketing techniques to encourage the purchase of environmentally-friendly products or applying such techniques more generally to create behavioral change even where there is no product involved. Market segmentation techniques, for example, are used to characterize different sectors of the target audience according to the motivations presumed to underlie their willingness to undertake behavioral change. As a result, messages are tailored to fit with the particular values dominant within different segments of the target audience—rather than reflecting more deeply on whether these are helpful values to be engaging.

Consequently, it is frequently asserted that campaigners should be indifferent to the motivations that underlie behavioral choices. Much as in the case of selling a product, they should "go with what works." Frequently, this may entail encouraging individuals to change their behavior for reasons of social status or financial self-interest rather than environmental benefit.[1]

The Failure of the Marketing Approach

Marketing approaches to creating behavioral change may be the most effective way of motivating specific change on a piecemeal basis. But the evidence presented in this report suggests that such approaches may actually serve to defer, or even undermine, pro-

---

[1] The arguments presented in "Weathercocks" have since been developed further. It is clear that there is a constellation of intrinsic values—not just environmental ones—which underpin systemic concern about environmental (and for that matter, social) problems.

spects for the more far-reaching and systemic behavioral changes that are needed.

There is little evidence that, in the course of encouraging individuals to adopt simple and painless behavioral changes, this will in turn motivate them to engage in more significant changes. The results of experiments examining the foot-in-the-door approach (the hope that individuals can be led up a virtuous ladder of ever more far-reaching behavioral changes) are fraught with contradictions. Current emphasis on "simple and painless steps" may be a distraction from the approaches that will be needed to create more systemic change. Such emphasis also deflects precious campaign and communication resources from alternative approaches.

Of course, this is not to argue that engaging in simple pro-environmental behaviors such as turning TVs off stand-by or switching from incandescent to compact fluorescent light bulbs is inherently wrong (*en masse*, these behavioral changes can clearly help). But it is to argue that such behaviors are the wrong focus for pro-environmental behavioral change strategies.

Worse, emphasis on the opportunities offered by green consumption distracts attention from the fundamental problems inherent to consumerism. This report reviews arguments that the consumption of *ever more* goods and services is an inherent aspect of consumerism and that the scale of environmental challenges we confront demands a systemic engagement with this problem. While alternative patterns of consumption (for example, car sharing or keeping and upgrading computers rather than replacing them) are important, these models cannot be properly disseminated, and seem unlikely to lead to change on the scale required, without first engaging the underlying motivations for consumerism.

Car sharing, for example, may not lead to net environmental benefits if the money that an individual saves by selling their own car and joining a car-share scheme is spent on buying into a time-share apartment in Spain. Treasuring objects for longer may not help either, if, rather than buying a new computer each year, a consumer upgrades their existing one and spends the money saved on another new electronic product.

This report also argues that, contrary to the assertions of proponents of marketing approaches, the *reasons* for adopting particular behavioral changes have very important implications for the energy and persistence with which these behaviors are pursued.

An individual might be less inclined to spend money saved by selling their car on an additional foreign holiday if they were motivated to part with their car for environmental reasons as opposed to economic incentives. Similarly, to the extent that specific pro-environmental behaviors may "spill-over" into other behaviors under some circumstances, such spill-over may be encouraged if initial behavioral changes are adopted for environmental reasons—as opposed, for example, to financial savings.[2]

Lessons from the Marketing Approach

Despite these criticisms, there are some lessons that should be drawn from marketing approaches to motivating pro-environmental behavioral change.

Proponents of the marketing approach recognize the importance of values in driving behavioral choices—even if they tend to argue that dominant values should be taken as "given." This is a crucial point. Firstly, it underscores the recognition that we should not expect information campaigns to create behavioral change. Secondly, it has an important bearing on our understanding of the gap between what people say and what they do.

It has been argued that it is futile to engage values and identity in the course of pursuing pro-environmental behavioral change, because of the so-called "attitude-behavior gap." This is the disparity between the importance that an individual may ascribe to environmental issues when interviewed and his or her actual behavioral patterns. This report draws a distinction, however, between attitudes and values and points to evidence which under-

---

[2] Clear experimental evidence, generated since "Weathercocks" was published, now supports this contention. See, for example, Evans et al. (2012), referenced on p.217.

scores the importance of engaging values and self-identity as a basis for motivating pro-environmental behavioral change.

Work on marketing approaches to motivating behavioral change also highlights the need to communicate with different people in different ways. This is crucial, but it says nothing about the effects that a communicator may be seeking to achieve with such communication. It need *not* imply that communications should be constrained to work with those motivations which currently dominate within a particular audience. Rather, it may be necessary to work to bring other, latent, motivations to the fore, while of course communicating with different people in different ways.[3]

Proponents of the marketing approach are also right to emphasize the importance of social context. Whether motivating people to buy a smart electricity meter or to join a local carbon rationing action group (CRAG), social norms and status will be critically important. But again, this need say nothing about the values upon which those norms are based.

Finally, the wider constraints on adopting new behavior are generally well recognized by proponents of the marketing approach—and this understanding is critically important. Any campaign to motivate individuals to join a car share scheme will meet with more limited success if these cars are located far from where the target audience live, and any campaign to motivate people to leave their cars at home and commute by train will meet with more limited success if the trains are over-crowded and don't run on time. But, crucially important as such concerns are, there is evidence that the willingness of people to suffer inconvenience and difficulty in engaging in pro-environmental behavior is related to their motivations for doing so. Values underpinning environmental behavior will be of critical importance both in motivating individuals to engage in pro-environmental behavior where such barriers

---

[3] There is actually a wealth of evidence, which "Weathercocks" did not review, that individuals who are more extrinsically oriented nonetheless hold intrinsic values to be more important than extrinsic values.

persist and in activating public demand for government intervention to remove these barriers.

An Alternative Approach

This report begins to build an alternative approach to motivating
pro-environmental behavioral change. This approach draws not on
analogies from marketing but rather from political strategy. It is
supported by recent work that underscores the importance of
framing a political project in terms of the values that underpin
this—rather than constantly molding this project to reflect the results of the latest focus-group research. Any successful movement,
it is argued, must be unequivocal in articulating what it stands
for. But of course, in itself, recognition of the importance of achieving consistency and clarity in the values that underpin environmental campaigning says nothing about *what* those values should
be.

Some argue that it will be most effective to frame environmental campaigns in terms of a set of individualistic or even materialistic values—for example, highlighting the personal benefits that
can accrue from more efficient energy use or the social status that
might be conferred by ownership of a hybrid car.

But this report presents evidence that appeals to individualism
are unlikely to be adequate. Research has found that many people
have a more 'inclusive' sense of self-identity—one that may include
closer identity with other people or with other people and nature.
These individuals thus tend to value others more in their behavioral choices, and research has repeatedly found that such people
tend to care more about environmental problems, favor environmental protection over economic growth, and engage in more pro-
environmental behavior. The issue of how such values are nur-
tured and "activated" is critically important.

There is also evidence that materialistic values cannot form
the basis for motivating systemic pro-environmental behavioral
changes. Importantly, we pursue our self-identity through the
products we buy—our material possessions come to define who we
see ourselves as being and who we want to be seen to be. This is a

sense of identity which the marketing industry has become adept at manipulating in order to motivate us to buy particular products as a means of further developing and confirming this identity. And of course, these same marketing techniques are increasingly used to sell "green" goods and services.

Individuals who engage in behavior in pursuit of "intrinsic goals" (of personal growth, emotional intimacy or community involvement) tend to be more highly motivated and more persistent in engaging in this behavior than individuals motivated by "extrinsic goals" (for example, of acquisition of material goods, financial success, image and social recognition). Moreover, more materialistic individuals tend to have higher ecological footprints.

This report presents evidence that motivations which are intrinsic are more likely to lead to pro-environmental behavior. Moreover, this effect is found to be particularly strong for more difficult environmental behaviors—those requiring greater effort.

Conversely, motivations that stem from external motivations (for example, a financial reward for behavior) or even what are called "internalized forms of external constraints" (these might include a sense of guilt or feelings related to self-esteem) are less likely to lead to pro-environmental behavior. This evidence raises critical questions about whether 'simple and painless steps' urged upon us for reasons of self-interest will contribute to motivating an individual to engage in more significant (and potentially inconvenient or costly) behavioral changes.

So it may be critically important that a campaign to motivate pro-environmental behavioral change should reflect, unequivocally, the values that underpin this campaign. Moreover, the nature of these values themselves may also be of critical importance.

Given the scale and urgency of the environmental challenges we confront, these are important assertions, and this report highlights a number of possible practical responses.

However, this report also highlights the relevance of this debate for the future of environmental organizations themselves. The enthusiasm of the private sector to embrace environmental imperatives has raised questions about the continued relevance of environmental organizations.

It may be that environmental organizations will indeed become side lined in the debate, unless they are prepared to reframe their contribution in terms of a set of values that are distinct from those identified with the private sector.

Many will still see the approach outlined in this report as unrealistic. But that perception is changing. Unfortunately, it is changing in part because as our understanding of the severity of the environmental challenges that confront us develops further, current strategies for engaging them seem increasingly inadequate. But WWF is also finding an increasing number of people, not easily pigeon-holed as environmentalists, who are nonetheless embracing a radical change agenda from within their respective sectors. The irony is that the mainstream environmental movement has yet to take on a leading role in responding to this challenge.

April 2008, London

Winning the Conversation: Framing and Moral Messaging in Environmental Campaigns

Matthew C. Nisbet

Ezra M. Markowitz

John E. Kotcher

Tom Crompton's "Weathercocks and Signposts" is an important call for new directions in environmental communication. As Crompton aptly argues, campaigns that narrowly seek to promote sustainable consumer choices and personal behavior may "serve as dangerous distractions from the serious business of getting in place policy frameworks that are sufficiently ambitious" (p. 13) in addressing climate change, energy insecurity, and related problems. As Crompton explains, environmental campaigns need to better assess different publics relevant to an issue, deploy a broader vocabulary in making the case for change, and enable a greater diversity of voices to express and demonstrate their support for action.

In this chapter, drawing on research from the social and behavioral sciences, we analyze campaign efforts in the United States to mobilize public demand for policy action on climate change, elaborating in detail on the themes and conclusions offered by Crompton. We describe methods for reframing climate change in ways that are more personally engaging, for creating a moral foundation that compels greater participation, for localizing the issue and switching policy focus, thereby diffusing political polarization, and for using opinion-leaders as community-level connectors and recruiters.

Public Participation: Processes and Barriers

Public participation relative to climate change, energy insecurity, and related environmental problems can take multiple forms, and these forms mirror how individuals participate in public life more generally. They include "political participation," defined as expressing political opinions and preferences to decision-makers and to peers while also recruiting others to become involved; "civic engagement," defined as working with others to collaboratively address complex problems and challenges in local communities and regions; and "political consumerism," which includes rewarding and punishing businesses and organizations for their products or practices and encouraging peers to do the same (Nisbet & Kotcher, 2009).

From a social change perspective, research suggests four major reasons why these forms of public participation matter to challenging the status quo on environmental problems, especially in relation to climate change. First, public participation rewards and punishes societal decision-makers (Verba, Schlozman, & Brady, 1995). As environmental groups admitted following the defeat in 2010 of U.S. cap and trade legislation, the lack of public pressure in support of the bill, especially in key states, contributed to defeat. "The community that tried to move a climate bill fundamentally lacks political power and doesn't have the ability to either deliver punishment or reward to members of Congress who don't vote for us," said Kathleen Welch, a Washington-based philanthropy adviser (Samuelsohn, 2011).

Second, public participation can build trust and collaboration. Not all forms of participation involve political activity. Research has shown that various forms of community involvement, such as church attendance or volunteering, build trust and social capital among participants (Putnam, 2000). If civic involvement focuses on defending and protecting local communities and people from climate change impacts, some participants may eventually be recruited into political activity on behalf of national policy action focused on curbing emissions or investing in new energy sources.

Third, public participation has the potential to improve the quality and perceived legitimacy of policy action. Experts and policymakers do not have all the answers, and nationally derived one-size-fits-all approaches to regional and community problems often fail. Members of the public offer their own forms of expertise based on professional training and personal experience that can and should inform policy approaches to climate change. Studies also show that when members of the public feel as if experts and decision-makers have genuinely listened to them, public participation helps diffuse polarization and builds consensus for eventual policy actions (Einsiedel, 2008).

Finally, public participation serves as a moral force in society. Public activism and expression can alter society's overall "quasi-statistical sense" of what the culture expects, accepts, and believes to be just. This is especially the case when those who are the most vocal are perceived by other individuals to be part of their "in group" and are able to frame the reasons for action in moral terms (Noelle-Neumann, 1995).

Pathways to Public Participation

Those individuals most likely to become involved in public life typically benefit from greater time, money, organizational and communication skills; are more politically interested, knowledgeable, opinion-intense, efficacious, and trusting; and receive a greater number of requests to become involved from peers and groups (Verba et al., 1995; Nie, Junn & Stehlik-Barry, 1996). However, social settings and information sources play an important role above and beyond these background factors and are especially important in the context of developing successful communication campaigns.

Social settings such as church, work, and volunteer groups serve as "networks of recruitment," where individuals receive requests to become involved on issues like climate change (Goidel & Nisbet, 2006). Studies show that the more requests a person receives in these types of settings, the higher the level of their participation (Verba et al., 1995). These social settings also indirectly provide the hard and soft resources needed to participate. Examples of hard resources include a space to meet and access to computers, photocopiers and phones. In terms of soft resources, Robert Putnam's work on social capital highlights the role of trust in institutions, group membership, and time spent interacting with others in one's community as primary motivators of civic participation (Putnam, 2000).

Church groups, for example, often rely on strong interpersonal bonds and norms of stewardship to encourage their members to participate in activities that they otherwise would be unlikely to engage in. These networks are further strengthened by the moral and political framing of issues by church leaders, the conversations that church-goers have with others, and information provided directly when at church (Goidel & Nisbet, 2006). Requests to participate and conversations that individuals have at work serve similar functions (Scheufele, Nisbet, & Brossard, 2003). At work, many individuals serve in leadership positions with this experience making them more effective as political participants and peer recruiters (Nie et al., 1996). Higher education also often translates into greater leisure time and membership in a greater diversity of community groups. In turn, these translate into a greater number of weak ties and connections (what Putnam refers to as "bridging" social capital), important resources for recruiting and relying on others to become involved on an issue (Putnam & Campbell, 2010).

Building on studies examining socio-economic background and social settings, communication research has examined the connections between these factors and forms of media, Internet, and mobile technology use as well as interpersonal conversations and social media interactions. This research finds that forms of media use and different types of conversations promote participation by providing "mobilizing information" that tells individuals how to get involved and who to contact and by defining issues in personally relevant ways (Campbell & Kwak, 2010; de Zúñiga et al., 2010; Eveland & Scheufele, 2000).

Psychological Barriers to Public Participation

If the basic mechanisms shaping public participation on an issue can be reduced to socio-economic status, recruitment through social settings, and mobilization through information sources, why hasn't public involvement on the issue been greater, given the amount of resources devoted to climate change communication in recent years?

To begin answering this fundamental question, researchers have identified and explored numerous psychological and social processes that shape how and to what extent individuals engage with climate change. Many of these mechanisms act as barriers to public participation, either by directly weakening motivation to participate or else indirectly by inhibiting perceptions of climate change as an important, real, and serious problem that requires action (Gifford, 2011). Communication and campaign efforts that do not take these factors into account and address them are unlikely to be successful.

Most Americans are still not yet actively engaged with climate change—cognitively, emotionally or behaviorally (Pew, 2010). Increasing and diversifying forms of public participation on climate change issues will involve overcoming behavioral inertia and individuals' deeply engrained and well-practiced habits, since participation relative to climate change, much less public affairs more generally, is an uncommon occurrence rather than a top-of-mind habit for most individuals.

Further depressing motivation to engage with climate change, many Americans' perceive, correctly, that the worst consequences of the problem will accrue to other people living far away both in time and space. Because people strongly discount events that are psychologically distant, such perceptions generally act to decrease motivation to participate (Trope, Liberman, & Wakslak, 2007). Moreover, because climate change may be a distant, abstract, probabilistic, scientific and personally "un-experiencable" phenomenon, people lack naturally strong emotional reactions to the issue (Weber, 2006). Much research in the behavioral sciences now points to the critical role that such affective processes play in shaping both our short- and long-term decision-making.

Climate change also involves uncertainty relative to its ultimate consequences, timing of events, and severity of outcomes. In turn, uncertainty tends to promote the belief that individuals do not hold a personal responsibility to respond, since people are generally predisposed to be overly optimistic about personal risks (Weinstein, 1980). As a result, many individuals are likely to interpret information about climate change in ways that allow them to maintain a positive outlook on life. When people do recognize the severity of the threats posed by climate change, they are likely to respond either by denying those consequences or else by feeling helpless to respond (Norgaard, 2011). Alternatively, attempts to educate people about the threats through fear-based appeals can actually increase skepticism about climate science when individuals hold a strong belief that the world is a just, fair, and stable place (Feinberg & Willer, 2011).

Feelings of inefficacy also pose a significant barrier to greater levels of public participation, especially when individuals are told that there is little that can be done to stop the worst outcomes of impending changes (a message that is increasingly coming from both climate skeptics *and* the most ardent climate advocates). Similarly, when individuals lack trust in existing institutions and actors do not have either the capacity to respond effectively or the motivation to do so, personal motivation to engage with the issue is dampened, since feelings of reciprocal sacrifice are important motivators of participation.

Political Barriers to Public Participation

As political leaders and activists have increasingly packaged almost every policy debate in terms of clearly defined ideological differences, party labels have become brand names, each standing for a distinct set of conservative or liberal positions. Partisan elite cues—as political scientists call them—cut down on the amount of information highly-educated and politically engaged Americans need to form an opinion about a topic, enabling strong Democrats

and Republicans to bundle their opinions in an ideologically consistent direction across issues. Politically attentive Democrats tend
to consistently take a strong liberal position on issues, and politically attentive Republicans tend to take an even stronger conservative position (Abramowitz, 2010). Over the past decade, as
elected officials and political leaders have diverged on climate
change, the public voices on the issue have tended to reflect a
deeply polarized minority, while moderate voices have become increasingly less visible, reinforcing policy gridlock, and creating
barriers to wider participation on the issue.

Strengthening the tendency to ideologically bundle their opinions, highly motivated partisans tailor their media consumption to
sources that reflect their ideological outlook (Mutz, 2006). On cable
television, Fox News, MSNBC and CNN offer viewers equally one-
sided presentations on the issue of climate change, though with
different slants. A majority of segments at Fox News dismiss the
need for action on climate change while a strong majority of segments at MSNBC and CNN strongly endorse reasons for action.
Predictably, among Republicans, heavier viewers of Fox are more
dismissive of climate change than their lighter-viewing counter-
parts. Yet those few Republicans who watch the other cable networks are more similar to Democrats in their views about the
issue. An implication is that it may be possible to engage Republicans on the issue of climate change if the issue is recontextualized
in a persuasive and compelling way, a topic we return to later in
the chapter (Feldman et al., 2012).

With Obama's election in 2008, among Republican and con-
servative-leaning Americans, skepticism of climate science became
a stand-in for opposition to a climate policy that Republican leaders and conservative commentators defined as damaging to the
economy. According to Patrick Moynihan and Gary Langer
(Moynihan, Langer, & Craighill, 2010), directors of survey research at Harvard University and ABC News respectively, during
this period, when Republicans in surveys increasingly answered
that they did not believe in man-made climate change, their an-

swer served as "a vehicle to express antipathy toward the solution, not to voice firm belief in the existence of the problem" (p. 15).

Consider also how this polarization process has shaped views on government investment in wind and solar energy sources. When President Obama took office in 2009, more than 80 percent of Republicans and Democrats favored government investment, reflecting consistent elite cues from leaders of both parties advocating on behalf of the promise of solar and wind. Yet as Obama began to make clean energy a major part of his domestic policy agenda, Republican leaders shifted their position, a shift reflected in discussion at conservative media outlets. The divergence in elite cues on the issue spiked in the wake of the Solyndra bankruptcy inquiry. By late 2011, support among Republicans had dropped to 53 percent, a 30 percent gap with Democrats (Pew, 2011).

Recent experimental work by Yale University's Dan Kahan and colleagues (2007) supports the conclusion that perceptions of climate science are policy dependent. In these studies, when conservatives read that the *solution* to climate change was more nuclear power, their skepticism of expert statements relative to climate change decreased and their support for policy responses increased. In contrast, when the solution to climate change was framed as stricter pollution controls, conservatives' acceptance of expert statements on the causes of climate change decreased, whereas that of liberals increased. A major implication is that effective public engagement on climate change will depend on the policy actions proposed, with some actions such as tax incentives for nuclear energy, government support for clean energy research, or proposals to defend and protect local communities against climate change impacts more likely to gain support and participation from both Democrats and Republicans.

Breaking Barriers: A Moral Foundation for Action

Historically, advocates and their campaigns have defined climate change as an environmental problem that threatens ecosystems and wildlife, often in remote polar regions or other countries. Similarly, calls for policy action and the public to become involved have been focused on the national and international level, defined relatively narrowly in terms of mitigating greenhouse gas emissions. As a consequence, outside a committed base of environmentalists and progressive activists, the public still lacks strong moral intuitions about climate change with appeals to participate lacking moral weight. Rather than identify the issue as one that involves considerations of right and wrong, blame and responsibility, many people understand climate change in scientific, technical, political and economic terms (Markowitz, 2012). If they do perceive a moral dimension to climate change, it is likely relative to care for the environment, an intuition that is easily overlooked in the context of economic concerns.

As Crompton urges, in the following section we discuss the need for environmental organizations to broaden and deepen their moral vocabulary, a shift that is likely to diversify and intensify forms of public participation. We describe the promise for reframing climate change in terms of public health and local-level actions that protect people and communities from climate change impacts.

Framing and Moral Intuitions

Frames are interpretive storylines that set a specific train of thought in motion, communicating why an issue might be a problem, who or what might be responsible for it, and what should be done. Framing a policy problem or issue endows certain dimensions of the complex issue with greater apparent relevance than they would have under an alternative frame. Audiences rely on frames to make sense of and discuss an issue; journalists use frames to craft interesting and appealing news reports; policymakers apply frames to define policy options and reach decisions; and, experts employ frames to simplify technical details and make them persuasive. As interpretative packages for an issue, frames are instantly translated by way of catch phrases (green jobs), metaphors (comparison to the Space Race), and visuals (construction worker retrofitting a house) (Nisbet, 2009; Scheufele, 1999).

Frames are often most effective when they appeal to morally relevant intuitions that are strongly held by an individual or segment of the public. "Part of what it means to be a partisan is that you have acquired the right set of intuitive reactions to hundreds of words and phrases," explains University of Virginia psychologist Jonathan Haidt (2012) in his best-selling book *The Righteous Mind.* "Within a few seconds or less of encountering phrases like pro-life or pro-choice, your intuition has already started to lean one way, and that lean influences what you think and do next" (p. 58). In his research, Haidt draws on surveys of tens of thousands of individuals to develop and validate a typology of six commonly held moral foundations, which include:

- *Harm/care*—concerns about the caring for and protection of others
- *Fairness/cheating*—concerns about treating others fairly and upholding justice
- *Loyalty/betrayal*—concerns about group membership and loyalty to one's nation and community
- *Authority/subversion*—concerns about legitimacy, leadership, and tradition
- *Liberty/oppression*—concerns about personal freedom and control by others
- *Sanctity/degradation*—concerns about purity, sanctity and contamination

Liberal advocates and environmental leaders tend to communicate about issues in ways that mostly activate the *moral foundations of harm/care, liberty/oppression,* and *fairness/cheating.* This reflects their own intuitive lens in how they make sense of issues and the vocabulary they share with like-minded colleagues, friends, and supporters. Yet to intensify and diversify public participation on climate change, a successful campaign needs to also appeal to a greater bandwidth of moral foundations and to be fluent in a variety of "moral languages."

From *An Inconvenient Truth* to Repower America

Historically, environmentalists and climate advocates have displayed a relatively narrow moral palette, speaking only a few moral languages in their efforts to intensify public participation or in some cases failing even to offer a moral argument for action. In the 2006 film *An Inconvenient Truth*, former vice president Al Gore dramatized the climate crisis for viewers by emphasizing specific environmental impacts, including hurricane devastation or famous cities or landmarks under water due to future sea-level rise. This framing strategy activated the *moral foundation of harm/care*, though much of the focus was on harm to nature or the environment rather than humans. The film, as captured in its title, had a heavy focus on holding industry and conservatives accountable for inaction (Nisbet, 2009), activating the moral foundations of *liberty/oppression* and *fairness/cheating*.

In 2008, Gore announced in a *60 Minutes* interview that he would embark on a three-year, $300 million "We Can Solve It" advertising campaign, designed "to recruit 10 million advocates to seek laws and policies that can cut greenhouse gases." Gore conceived of the WE advertising campaign as addressing the problem of donor confusion and overlapping competitive strategies among environmental groups. "Everyone is faced with a continuing struggle of funding from donors. And the more this issue rises, the more it is used in all the appeals, and that's fine in a way," Gore told environmental leaders in 2007. "But the message is chewed up and ends up not feeding the growth of a truly mass movement" (Pooley, 2010, p. 25).

The WE campaign marked a shift in communication strategy for Gore. He would no longer be at the forefront of media messaging, and the framing focus would be recast to focus on national unity. In the first TV ads framed around the theme of "We Can Solve the Climate Crisis," the issue was characterized as a solvable and shared moral challenge, activating the *moral foundation of loyalty*. The ads paired unlikely spokespeople such as Speaker of the House Nancy Pelosi with former Speaker of the House Newt Gingrich.[4] In ads titled "America Does Not Wait," action on global warming was compared to the U.S. Civil Rights Movement, the

---

[4] The Pelosi ad is on YouTube at: http://youtu.be/VaZFfQKWX54.

United States' role in aiding allies in World War II, and the recovery from the Great Depression.[5]

In 2009 and 2010, as cap and trade legislation was debated, Gore's campaign was recast again, this time around the theme of "Repower America," as TV ads featured actors as ranchers, construction workers, and autoworkers, stressing the economic benefits of action on climate change.[6] The Environmental Defense Fund—as the chief group behind the cap and trade policy approach—ran similar ads in Midwest swing districts and cities. The ads were framed around the theme of "cap = jobs." Narrated by Braddock, Pennsylvania mayor John Fetterman, the ads featured images of tough steel workers and told viewers: "We need a cap on carbon pollution. It'll create jobs making things like solar panels and wind turbines. There's two hundred tons of steel in wind turbine. You guys can handle that, right?" The ad then flashed the slogan: "Carbon caps = Hard hats."[7]

A Failure to Mobilize. Yet at a time when public participation was needed the most, the Repower America and "cap = jobs" campaigns lacked a strong moral foundation. The campaigns offered the promise of economic benefits but did not build a case for why we should act and why we have a responsibility to do so. The emphasis on economic benefits in the context of the recession also turned the debate into "some economic benefits" as claimed by greens versus "dramatic economic costs" as claimed by opponents, a balance that, given the economic context, favored the opposition.

The table below summarizes the moral foundations that were activated by various recent climate change campaigns. In 2006 and 2007, *An Inconvenient Truth* (ACT) focused on moral intuitions that engaged a liberal base, and when *harm/care* was activated, the focus was more on the environment than human welfare. In 2008, the We Can Solve It (WE) campaign expanded to

---

[5] The "America Does Not Wait" ad is on YouTube at: http://youtu.be/1fMgTD92nYk

[6] The Repower America ad is on YouTube at: http://www.youtube.com/watch?v=GlQc9Kj15NM&feature=relmfu

[7] The Cap = Jobs ad is on YouTube at: http://www.youtube.com/watch?v=IaJE1sjqEt4

include a focus on loyalty. In 2009 and 2010, as cap and trade was debated, the Repower America and cap = jobs campaigns lacked an obvious moral call to action, focusing instead narrowly on economic benefits.

Table 1. Moral Foundations Activated by Climate Campaigns, 2006–2010

Campaign	Held by Liberals and Conservatives		
	Harm/ Care	Fairness/ cheating	Liberty/ oppression
ACT	***	✔	✔
WE	***	–	–
Repower	–	–	–
Cap = Jobs	–	–	–
	Held Mostly by Conservatives		
	Loyalty/ betrayal	Authority/ subversion	Sanctity/ degradation
ACT	–	–	–
WE	✔	–	–
Repower	–	–	–
Cap = Jobs	–	–	–

*Note: *** Predominant focus was on harm to environment rather than human health or well-being.*

The lack of moral urgency and context is reflected in public opinion surveys. In January 2010, even after the House had passed cap and trade legislation and the Senate had taken up debate, 60 percent of Americans said they had not heard of cap and trade. Among the 40 percent who had, when asked what came to mind about the bill, 15 percent said nothing. Even among the 13 percent of Americans who could be categorized as genuinely "Alarmed" by climate change, when cap and trade legislation was specifically explained for them, their support remained soft. Moreover, among this Alarmed group, only 30 percent answered that they had engaged in any form of political action such as contacting an elected official (Revkin, 2011).

Local Action to Protect Innocents

In response to the failure of the climate bill, environmental leaders called for new approaches to communication, asserting that with national policy stalled, now was the time to invest in building networks and partnerships in the Midwest and other regions. "We will have to reach out to new partners, make new allies and engage new constituencies," wrote the Environmental Defense Fund's Fred Krupp (2010) at *The Huffington Post*. Moving forward, several points of evidence suggest that campaigns that reframe climate change in terms of public health and the harm to innocents may be the type of innovation needed to intensify and diversify public participation.

The public health frame stresses climate change's potential to increase the incidence of infectious diseases, asthma, allergies, heat stroke, and other salient health problems, especially among the most vulnerable populations: the elderly and children. In the process, the public health frame makes climate change personally relevant to new audiences by connecting the issue to health problems that are already familiar and perceived as important (Nisbet, 2009). The public health frame may also be especially effective because it can link climate change with humans' deeply-engrained disgust response, a powerful motivator of ameliorative action (Pizzaro, Inbar & Helion, 2011). The frame also shifts the geographic location of impacts, replacing visuals of remote Arctic regions, animals, and peoples with more socially proximate neighbors and places across local communities and cities. Coverage at local television news outlets and specialized urban media is also generated. In each case, the public health frame activates the moral foundation of *harm/care* with a focus on people, rather than the environment (Nisbet, 2009).

Research suggests that by focusing communication on the health risks of climate change and the health benefits of local-level action, even those doubtful or dismissive of climate change support local policy actions, a starting point for building support for eventual national policy (Maibach et al., 2010). Other research suggests that recasting action in terms of benefits to public health activates positive emotions of hope among those disengaged on the issue while diffusing common reactions of anger among those otherwise dismissive (Myers, Nisbet, Maibach, & Leiserowitz, 2012).

Focusing on the public health risks to innocents also binds together support for action while stigmatizing opponents. One of the critical turning points in the effort to combat smoking in the U.S. involved morally stigmatizing tobacco companies and smokers by shifting the focus from personal health effects to the negative effects of smoking on innocent bystanders, namely children. Once smoking became more than just a personal consumption choice or an industry regulation issue, anti-smoking advocates were able to catalyze greater public demand for bans on smoking in public spaces, cigarette taxes, and restrictions on tobacco sales and marketing (Brandt, 2007).

Efforts to protect and defend people and communities are also easily localized. State and municipal governments have greater control, responsibility, and authority over climate change adaptation-related policy actions. In addition, recruiting Americans to protect their neighbors and defend their communities against climate impacts naturally lends itself to forms of civic participation and community volunteering. In these cases, because of the localization of the issue and the non-political nature of participation, barriers related to polarization may be more easily overcome. Moreover, once community members from differing political backgrounds join together to achieve a broadly inspiring goal like protecting people and a local way of life, then the networks of trust and collaboration formed can be used to move this diverse segment toward cooperation in pursuit of national policy goals.

A leading example of the public health message strategy is a TV ad in the 2012 "Beyond Coal" campaign sponsored by the Sierra Club, Natural Resources Defense Council, and the American Lung Association. The ads emphasize the risks to innocents while stigmatizing opponents to action. Depicting asthmatic kids walking the halls of Congress with inhalers, the TV ad tells viewers: "If every lobbyist in Congress were a severely asthmatic kid, then maybe lobbyists wouldn't be telling Congress to gut clean air

standards...support the President's and the EPA's clean air standards, and support our kids."[8] In combination with these TV ads,
using grassroots organizing strategies, the Beyond Coal campaign
has mobilized volunteers around efforts to shut down local coal
power plants, actions that more easily translate into local benefits
such as cleaner air and water and a transition toward innovative
industries and energy sources that create higher quality jobs
(Hertsgaard, 2012).

Breaking Barriers: Winning the Conversation

When Gore re-launched his Repower America campaign in Fall
2011 as the Climate Reality Project, he argued that people "needed
to win the conversation" on climate change much like the public
did on racism and civil rights during the 1960s. Across the next
two decades, it became publicly unacceptable to make racist comments and to engage in racist acts. The same, argued Gore, needs
to happen relative to statements denying the reality of climate
change or action opposing social change (Huffington Post, 2011).

Research across several different disciplines has consistently
shown that individuals frequently monitor their social environment and conform their opinions and decisions to the perceived
majority norm. Research also shows that people have a "faulty
quasi-statistical" sense in that they are often not very good at de-
termining what the true opinion of the majority might be (Noelle-
Neumann, 1995). Instead they rely on media portrayals and are
often persuaded by the most vocal group or readily memorable example. Importantly, what individuals often pay closest attention to
is the perceived opinion among people like them, in other words
their direct reference group (Scheufele & Moy, 2000).

In this context, Gore's principle is correct, but it will take more
than environmentalists evoking climate change as a moral imperative to widen and diversify the issue public on climate change and
to intensify public demand for action. Instead, the identification
and recruitment of opinion-leaders across reference groups are

---

[8] The Beyond Coal ad can be watch at YouTube: http://youtu.be/GbrNZjRqfmQ

needed, voices that can activate specific moral foundations that compel feelings of responsibility and obligation. Opinion-leaders are everyday individuals who have a stronger motivation for public affairs content or information specific to climate change and who have a special ability as a trusted source to share that information with others. Opinion-leaders rarely hold formal positions of authority and instead prove influential by way of their greater attention to a topic, their knowledge, and their strength of personality and experience in serving as a central go-between for information among their large network of core and loose ties.

As a combination of these traits and behaviors, opinion-leaders not only help draw the attention of others to a particular issue, action, or consumer choice but, perhaps most importantly, signal how others should in turn respond or act. This influence may occur by giving advice and recommendations, by serving as a role model that others can imitate, by persuading or convincing others, or by way of contagion, a process where ideas or behaviors are spread with the initiator and the recipient unaware of any intentional attempt at influence (Nisbet & Kotcher, 2009; Shenk & Dobler, 2002).

Opinion-Leaders as Recruiters and Connectors

For a number of reasons, opinion-leader identification, recruitment, and coordination are likely to be central to climate change campaigns, especially those seeking to spark greater levels of local-level participation.

First, at church, work, or in other face-to-face social settings, opinion-leaders are typically the major source originators of requests to get involved and participate among their peers and co-workers. By modeling civic and political participation on climate change, opinion-leaders also increase perceptions of efficacy among their peers, i.e., that they too can successfully become involved. They also signal that civic commitment to climate change is socially normative and desirable. Similar modeling takes place surrounding consumer choices and the adoption of new energy-related behaviors.

Second, opinion-leaders also bridge online audience gaps by passing on and sharing news and information about climate change that their peers would otherwise never be exposed to. This is especially important in coordination with major climate change-related focusing events such as the release of a new government report; a local event or political decision; a pending national decision; or within the context of a primary or general election.

Third, through conversations and social media, opinion-leaders can additionally serve as direct peer-educators, informing and instructing their friends and family on how to engage in different forms of participation. They can also break down competency gaps in the use of digital technology by modeling the use of mobile and hand-held devices or by teaching others how and where to access information and digital tools (Nisbet & Kotcher, 2009).

Previous research suggests three types of opinion-leaders that are likely to be important depending on the goal and focus of a climate change campaign. Survey scales have been developed to reliably and validly identify these categories of opinion-leaders. Shortened versions can be embedded in email or web surveys by organizations.

Climate Change-Specific Opinion Leaders. These individuals are unique in that they pay very close attention to news and information about climate change and energy and otherwise share similar traits with "influentials." These types of opinion-leaders are best suited to recruiting individuals from among segments of the public already predisposed to be alarmed or strongly concerned by the issue of climate change. Many activists in past and current climate change campaigns exhibit strong issue-specific opinion-leader traits (Nisbet & Kotcher, 2009).

Influentials or Public Affairs Generalists. To widen the appeal and recruitment potential of a climate change campaign, so-called "influentials" should also be a strong focus of recruitment. These opinion-leading generalists track public affairs news and issues more closely, have overall higher levels of civic involvement, social capital, and political participation as measured by group membership and involvement and tend to score higher in terms of strength of personality (Weimann, 1994). These types of opinion-leaders are ideally suited to recruiting people to participate politically or civically from among audience segments normally disengaged or doubtful about the importance of climate change and less trusting of environmental groups.

Market Mavens and Communicative Adopters. Consumer behavior research has identified "market mavens" as a special class of consumers who take pleasure in shopping, follow closely the release of new products as well as sales and discounts, and enjoy sharing this information with others (Feick & Price, 1987). Communicative adopters are not only generally first generation purchasers of new products and technologies, but they also evangelize and recommend the product to others (Venkatraman, 1989). Across segments of the public, these opinion-leaders are likely to be especially important for promoting forms of political consumerism, rewarding and punishing companies.

Examples of Opinion-Leader Campaigns

Several past campaigns offer useful examples of how opinion-leaders were integrated into recruitment and mobilization strategies.

Bush 2004 Campaign. In this example, campaign organizers sent an email questionnaire to their national list of seven million volunteers, asking four specific questions about how willing volunteers were to write letters to the editor, talk to others about politics, forward emails, or attend public meetings. Based on answers to these questions, the Bush team segmented out two million opinion-leaders.

Contacted on a weekly basis by email and phone, these two million self-designated opinion-leaders were asked to talk up the campaign to friends, write letters to the editor, call in to local radio programs, or attend public meetings staying on message at all times with nationally coordinated talking points. These supporters were used as grassroots information brokers, passing on interpersonally to fellow citizens the themes featured in political ads, news coverage, and in presidential stump speeches.

The Bush campaign reinforced their volunteers' commitment by providing personal access to VIPs such as a local Congressman or national campaign spokespeople. "Life-targeting" databases also allowed the campaign to match up their opinion-leaders with local voters who shared similar consumer preferences and product tastes, correlates that the campaign would use as proxy measures for predicting the effect of issue specific appeals. For example, an opinion-leader tagged as a "terrorism moderate" would be asked to call other "terrorism moderates" living in his/her surrounding county (Fournier, Sosnick, Dowd, 2006; Nisbet & Kotcher, 2009).

The Bush campaign specifically targeted churches as recruitment contexts, sending recruited opinion-leaders door-to-door in their surrounding neighborhoods. As part of this "friendship Evangelicalism" strategy, volunteers were simply asked to tell their neighbors why they backed Bush—to, in effect, witness their support for the president. Neighbors would then be invited to a house party of several dozen others from the community to watch a video about Bush, to have dinner or drinks, and to socialize. In contrast to the opinion-leader strategy of the Bush campaign, the Kerry campaign and aligned Democratic groups bused in paid canvassers to battleground states and districts, offering scripted and targeted appeals to voters that were often ineffective (Bishop, 2008, p. 264).

Obama 2008 Campaign. The My.BarackObama.com platform, launched in 2007, was a Facebook-like site that not only helped the campaign communicate with supporters and raise money, but it was also designed to help supporters connect with one another and organize events in their local community. Perhaps the most innovative strategy for combining digital tools and face-to-face opinion-leadership was the campaign's creation of an iPhone application. The interface organized friends and contacts by key battleground states, encouraged users to call their friends on behalf of Obama; provided information on local events; and included videos and issue backgrounders that users could reference or show during face-to-face conversations with friends. The application also provided feedback data to the campaign, such as the number of phone calls successfully completed (Shankland, 2008; Nisbet & Kotcher, 2009).

In training opinion-leaders, the goal was not to have them tell Obama's story to others but to persuasively narrate their own personal stories about why the campaign mattered to them. In designing a movement-themed campaign, the Obama team applied a philosophy to opinion-leaders that created a "structure that allows all members of the team to make [a] campaign his or her own." The Obama team believed that a campaign must go beyond "simply assigning volunteers to tasks to instead allow well-trained supporters to have ownership within the campaign (Berman, 2010, p. 124).

The We Can Solve It and Repower America Campaigns. Most of the TV advertising in the We Can Solve It and Repower America campaigns asked audiences to visit the campaign's web site, the main platform for putting into action recruited opinion-leaders. The major "ask" at the site was for visitors to sign up to be part of the campaign's action e-mail list so that "your voice can be heard." Immediately, for visitors, the most visually prominent feature of the site was the pop-up projection of a video of a volunteer Power-Point presenter from Gore's Climate Project, telling visitors in his or her own words why they need to get involved and/or explaining a feature of the site. Also prominent on the front page of the site was statistical information on the number of people to date "who want to be part of the solution" on climate change (Nisbet & Kotcher, 2009).

The WE and Repower America sites also featured a social networking component like Facebook, where visitors could create a profile, friend other people, write blog entries and letters to the editor, create groups, and attempt to organize local events in their community. These action alerts were coordinated with either a major vote in Congress, a major speech by Gore, or, for example, the launch of a new commercial during the August 2008 Olympics broadcast. As an incentive for contacting other citizens, participants who successfully encouraged 40 friends to sign up through word of mouth, forwarded emails and/or other social media actions

were named a "WE leader" and given "access to special information." The campaigns also launched their own Facebook application, where participants (referred to as Climate Champions) who
signed up fellow Facebook friends could earn points that donors
would then match as financial contributions to the campaign
(Nisbet & Kotcher, 2009).

Face to Face Versus Online?

The We Can Solve It and Repower America campaign were likely
limited because of an almost exclusive emphasis on online interaction and influence. Relying on digital connections and recruitment
is appealing because of the relative ease in which organizers can
develop metrics to measure success. Yet ease in tracking data does
not equate to effectiveness, and we urge caution in over-relying on
digital networks, especially in place of face-to-face influence. There
are several trade-offs and problems with relying too heavily on social media forms of opinion-leadership.

First, surveys show that Americans still prefer their recommendations via verbal interaction, and there still does not exist
strong research on whether or not the self-selection biases of the
Web can be overcome within digital networks. Moreover, to date,
by all accounts the success of the digital organizing efforts of the
WE campaign has been relatively modest at best. Second, with
strong selectivity bias online, exclusively relying on digital interaction might also result in ideological reinforcement and intensifica-
tion of beliefs about climate change, which may eventually limit
the willingness of recruited opinion leaders to compromise on
pragmatic policy solutions.

Third, if the "weak ties" of digital interactions lack the
strength of traditional opinion leader influence, then time and effort spent online by digital opinion-leaders may be far less effective
than traditional face-to-face influence. The danger of relying too
heavily on digital organizing is that it might create a false sense of
efficacy among participants, with activists believing they are making a difference on climate change, when impact may be limited at
best (Nisbet & Kotcher, 2009).

Given these many dimensions of online influence, the goal for both researchers and practitioners is to figure out under what conditions or with which demographic segments digital opinion-leaders can be effectively used on climate change campaigns and in which ways online interactions can build on real-world ties. Combining digital organizing with face-to-face interaction by using hand-held devices such as the iPhone, as was done in the Obama campaign, is a strategy that future climate change efforts should explore.

Conclusion

The call for new directions in Crompton's "Weathercocks" report is consistent with principles and conclusions derived from research in the social and behavioral sciences. In this chapter, we have elaborated on several specific areas where research points to additional insights and strategies. A first step for environmental organizations and their allies is to recognize the function and importance of different forms of public participation in creating the demand for societal change. A second step is to focus more specifically on framing the relevance of climate change and related challenges in a manner that activates a broader palette of moral intuitions, tailoring these appeals to specific segments of the American public.

Particularly promising are strategies that redefine climate change in terms of public health, the harm to innocents, and local level actions that can protect and defend people and communities. By focusing on these localized, incremental policy actions, networks of trust and collaboration are forged that can then translate into forms of participation that advocate on behalf of national policy action. Also highly relevant to successful campaigns are sophisticated and well-resourced strategies that employ opinion-leaders across audiences, empowering opinion-leaders to serve as trusted communicators, connectors, and recruiters.

However, formative research and ongoing evaluation are likely needed to refine and implement these strategies in ways that are useful to campaign professionals and organizers. Most notably, research on framing and climate change remains inadequate and limited. To date, the only framing approach to climate change that has been examined in a rigorous way has been work on public health appeals. In addition, formal research does not exist on how different framing strategies map to specific moral foundations and how this connection might apply differentially across segments of the public.

Research is also needed to develop efficient, valid, and trust-worthy methods for identifying different types of opinion-leaders specific to different forms of public participation, along with methods for training and using them effectively in campaign work. The internet has given campaigns new opportunities to collect massive amounts of data about their supporters that could be useful to opinion leader campaigns. For instance, once supporters register as a volunteer on the Obama 2012 campaign website, the campaign reserved the right to collect information about how volunteers used their website, such as what they click on and which pages they view; data about how they interact with campaign email messages; and personal information they submit as part of blog comments, interactive forums or contests and games on the campaign's websites. The growing use of Facebook authentication to let supporters log in to campaign websites also gives campaigns the ability to access a supporter's name, profile picture, gender, networks, list of friends and any other information they have made public (Beckett, 2012). While such data-mining techniques could prove beneficial, they also raise questions about privacy. And as we noted earlier, any online strategies need to be paired with face-to-face organizing and recruitment. In addition to examining the utility of this information to opinion leader campaigns, future research should also investigate the potential for this kind of data collection to compromise the trust of valuable campaign supporters.

References

Abramowitz, A. (2010). The disappearing center: Engaged citizens, polarization, and American democracy. New Haven: Yale University Press.

Beckett, L. (2012, Mar. 27). Three things we don't know about Obama's massive voter database. *ProPublica.*

Berman, A. (2010). Herding donkeys: The fight to rebuild the Democratic Party and reshape American politics. New York: Farrar, Straus, & Giroux. P. 124.

Bishop, B. (2008). *The big sort: Why the clustering of like minded America is tearing us apart.* New York: Houghton Mifflin Company.

Brandt, A.M. (2007). *The cigarette century: The rise, fall, and deadly persistence of the product that defined America.* New York: Basic Books.

Campbell, S.W., & Kwak, N. (2010), Mobile communication and civic life: Linking patterns of use to civic and political engagement. *Journal of Communication, 60,* 536–555.

de Zúñiga, H.G., Veenstra, A., Vraga, E., & Shah, D. (2010). Digital democracy: Reimagining pathways to political participation. *Journal of Information Technology & Politics, 7*(1), 36–51.

Einsiedel, E. (2008). Public engagement and dialogue: A research review. In M. Bucchi & B. Smart (Eds.), *Handbook of public communication on science and technology* (pp.173–184). London: Routledge.

Eveland, W. P., & Scheufele, D. A. (2000). Connecting news media use with gaps in knowledge and participation. *Political Communication, 17*(3), 215–237.

Feick, L.F., & Price, L.L. (1987). The market maven: A diffuser of marketplace information. *Journal of Marketing, 51*(1), 83–97.

Feinberg, M., & Willer, R. (2011). Apocalypse soon? Dire messages reduce belief in global warming by contradicting just world beliefs. *Psychological Science, 22,* 34–38.

Feldman, L., Maibach, E.W., Roser-Renouf, C., & Leiserowitz, A. (2012). Climate on cable. *International Journal of Press/Politics, 17*(1), 3–31.

Fournier, R., Sosnick, D., & Dowd, M. (2006). *Applebee's America: How successful political, business, and religious leaders connect with the new American community.* New York: Simon & Schuster.

Gifford, R. (2011). The dragons of inaction: Psychological barriers that limit climate change mitigation and adaptation. *American Psychologist, 66,* 290–302.

Goidel, K., & Nisbet, M.C. (2006). Exploring the roots of public participation in the controversy over stem cell research and cloning. *Political Behavior, 28*(2), 175–192.

Haidt, J. (2012). *The righteous mind: Why good people are divided by politics and religion.* New York: Pantheon Books.

Hertsgaard, M. (2012, Apr. 2). How a grassroots rebellion won the nation's biggest climate victory. *Mother Jones.*

Huffington Post (2011, Oct. 30). Al Gore on climate change deniers. *Huffington Post*. Available at: http://www.huffingtonpost.com/2011/08/30/gore-climate-change-deniers_n_940802.html

Kahan, D., Braman, D., Slovic, P., Gastil, J., & Cohen, G. (2007). The second national risk and culture study: Making sense of—and making progress in—the American culture war of fact (October 3, 2007). GWU Legal Studies Research Paper No. 370; Yale Law School, Public Law Working Paper No. 154; GWU Law School Public Law Research Paper No. 370; Harvard Law School Program on Risk Regulation Research Paper No. 08-26. Available at SSRN: http://ssrn.com/abstract=1017189

Krupp, F. (2010, Nov. 16). The new path forward on climate change. *The Huffington Post*. Available at: http://www.huffingtonpost.com/fred-krupp/the-new-path-forward-on-c_b_784182.html.

Maibach, E., Nisbet, M.C., Baldwin, P., Akerlof, K., & Diao, G. (2010). Reframing climate change as a public health issue: An exploratory study of public reactions. *BMC Public Health, 10*, 299.

Markowitz, E.M. (2012). Is climate change an ethical issue? Examining young adults' beliefs about climate and morality. *Climatic Change*. Advanced online publication. doi: 0.1007/s10584-012-0422-8

McLeod, J.M., Scheufele, D.A., & Moy, P. (1999). Community, communication, and participation: The role of mass media and interpersonal discussion in local political participation. *Political Communication, 16(3)*, 315–336.

Moynihan, P., Langer, G., & Craighill, P. (2010). An end to a means: Partisanship, policy preferences and global warming. Paper presented at the annual meetings of the American Association for Public Opinion Research (AAPOR).

Mutz, D. (2006). How the mass media divide us. In P. Nivola & D.W. Brady (eds.), *Red and blue nation?* Vol. 1. (pp. 223–263). Washington, DC: The Brookings Institution.

Myers, T., Nisbet, M.C., Maibach, E., & Leiserowitz, A. (2012). Hope or anger? Framing and emotion in the climate change debate. *Climatic Change, 113*, 1105–1112.

Nie, N. H., Junn, J., & Stehlik-Barry, K. (1996). *Education and democratic citizenship in America*. Chicago, IL: University of Chicago Press.

Nisbet, M.C. (2009). Communicating climate change: Why frames matter to public engagement. *Environment, 51*(2), 12–23.

Nisbet, M.C., & Kotcher, J. (2009). A two step flow of influence? Opinion-leader campaigns on climate change. *Science Communication, 30*, 328–354.

Noelle-Neumann, E. (1995). Public opinion and rationality. In T. L. Glasser & C. T. Salmon (Eds.), *Public opinion and the communication of consent* (pp. 33–54). New York: The Guilford Press.

Norgaard, K. (2011). *Living in denial: Climate change, emotions and everyday life*. Cambridge: MIT Press.

Pew Research Center for the People and the Press (2010). *Economy dominates public's agenda, dims hopes for the future.* Available at: http://www.people-press.org/2011/01/20/economy-dominates-publics-agenda-dims-hopes-for-the-future/

Pew Research Center for the People and the Press (2011, Nov.) *Partisan divide over alternative energy widens.* Available at: http://pewresearch.org/ pubs/ 2129/alternative-energy-solar-technology-nuclear-power-offshore-drilling.

Pizzaro, D., Inbar, Y., & Helion, C. (2011). On disgust and moral judgment. *Emotion Review, 3*(3), 267–268.

Pooley, E. (2010). *The climate war: True believers, power brokers, and the fight to save the earth.* New York: Hyperion Books.

Putnam, R.D. (1995). Bowling alone: America's declining social capital. *Journal of Democracy, 6*, 65–78.

Putnam, R. (2000). *Bowling alone: The collapse and revival of American community.* New York: Simon and Schuster.

Putnam, R.D., & Campbell, D.E. (2010). American grace: How religion divides and unites us. New York: Simon and Schuster.

Revkin, A. (2011, April 25). Beyond the climate blame game. Dot Earth Blog. The New York Times.com.

Samuelsohn, D. (2011, May 31). Green donors taking time to soul search. *Politico.* http://www.politico.com/news/stories/0511/55980.html

Schenk, M., & Dobler, T. (2002). Towards a theory of campaigns: The role of opinion-leaders. In H. Klingerman & Rommele, A. (Eds)., *Public information campaigns & opinion research.* London: Sage Publications, pp. 37–51.

Scheufele, D. A. (1999). Framing as a theory of media effects. *Journal of Communication, 49*(1), 103-122.

Scheufele, D.A., & Moy, P. (2000). Twenty-five years of the spiral of silence: A conceptual review and empirical outlook. *International Journal of Public Opinion Research, 12*(1), 3–28.

Scheufele, D.A., Nisbet, M.C., & Brossard, D. (2003). Pathways to participation? Religion, communication contexts, and mass media. *International Journal of Public Opinion Research, 15*(3): 300–324.

Shankland, S. (2008, Oct. 2). Obama releases iPhone recruiting, campaign tool. CNET news. Available at http://news.cnet.com/8301-13578_3-10056519-38.html.

Trope, Y., Liberman, N., Wakslak, C. (2007). Construal levels and psychological distance: Effects on representation, prediction, evaluation and behavior. *Journal of Consumer Psychology, 17*, 83–95.

Venkatraman, M.P. (1989). Opinion leaders, adopters, and communicative adopters: A role analysis. *Psychology and Marketing, 6*(1), 51–68.

Verba, S., Schlozman, K.L., Brady, H.E. (1995). *Voice and equality: Civic volunteerism in American politics.* Cambridge, MA: Harvard University Press.

Weber, E.U. (2006). Experience-based and description-based perceptions of long-term risk: Why global warming does not scare us (yet). *Climatic Change, 77,* 103–120.

Weimann, G. (1994). *The influential: People who influence people.* Albany, NY: State University of New York Press.

Weinstein, N. D. (1980). Unrealistic optimism about future life events. *Journal of Personality and Social Psychology, 39,* 806–820.

CHAPTER THREE

Biofuels and the Law of Unintended Consequences

Charles T. Salmon

Laleah Fernandez

In "Weathercocks and Signposts," Dr. Thomas Crompton identifies several important unintended consequences of social marketing interventions designed to promote pro-environmental causes. For example, he argues that initiatives designed to encourage citizens to take "simple and painless steps" as a means of promoting environmental interests may in fact make much-needed (and collectively painful) systemic change more difficult to institute. As well, he notes that financial incentives for consumers to "buy green" may not only be ineffective, they may be counter-productive as a result of their emphasis on consumerism and its concomitant emphasis on the production and consumption of more goods and services.

Recognition that social marketing interventions and other public policy initiatives may lead to unintended as well as intended consequences is essential to analyzing the manifest and latent implications of policy options. This chapter offers a case study of how the notion of unintended consequences can provide an analytic framework for studying public and political discourse about ethanol, a hybrid fuel composed of gasoline and biomass. In the course of this analysis, special attention is paid to how ethanol initially was framed by its proponents and subsequently reframed by its opponents, and the role that unintended consequences played in this reframing.

The Law of Unintended Consequences and Social Policy

Since the seventeenth century physicists and philosophers alike have posited that actions have consequences that are not necessarily intended and in some case not desired. The greatest physicist of his day, Sir Isaac Newton, articulated three laws of motion, the third of which posited that "to every action, there is always an equal and opposite reaction." In other words, force applied in one direction inevitably induces change in an unintended direction as well. Meanwhile, a contemporary of his, the noted political philosopher John Locke (1692), warned that a proposed cut in interest rates would not benefit borrowers as was intended but rather that it would redistribute wealth away from "widows, orphans and all those who have their estates in money."

Over the centuries, a number of noteworthy scholars similarly considered this topic in their writings: Machiavelli, Vico, Adam Smith, Karl Marx, Engels, Wundt, Pareto, Max Weber, Graham Wallas, Cooley, Sorokin, Gini, Chapin, von Schelting, among others (Merton, 1936, p. 2). Yet it wasn't until the twentieth century and the work of Robert Merton that the concept of unintended consequences was formally defined and its dimensions for study formally articulated.

Merton used the terms "manifest" and "latent" functions to differentiate between, respectively, the conscious motivations behind a socio-political action and the action's objective consequences. For example, the Clean Air Act of 1970 resulted in a number of utility companies building tall chimneys and smokestacks in order to fulfill the manifest function (and conscious motivation) of reducing pollution in the local vicinity. However, these tall chimneys merely dispersed pollution over a much wider geographic area and higher in the atmosphere, thereby resulting in a latent function (and objective consequence) of causing more severe environmental damage than the original problem that had been "solved" with the construction of taller chimneys. Similarly, familiar communication campaigns and improved firefighting techniques designed to prevent forest fires may have served the manifest function of reducing the incidence of such fires. Yet at the same time, these efforts also resulted in the latent function of forest fires occurring with greater intensity and causing more damage due to the accumulation of vegetation growth not cleared out periodically by naturally occurring fires. Indeed, across the entire spectrum of policy issues and of political systems, it is rare to find policy actions that have not engendered unintended consequences of one sort or another, often of equal or greater magnitude than the original problem being "remedied." It is for this reason that the inevitability of unintended consequences arising from new governmental regulations and programs has achieved status as a veritable "law" of public policy (Cho & Salmon, 2007; Salmon, 1992).

In explaining the prevalence of unintended consequences of policy actions, Merton focused on five key factors. First, policy makers often lack knowledge of the many details and facts necessary to make decisions with sufficient precision to narrowly induce change without accompanying unintended consequences. Second, errors happen in conceptualization, prediction, or implementation, and these errors spawn consequences. Third, the "imperious immediacy of interests" stemming from a preoccupation with solving one problem often prevents open-minded consideration of possible consequences that may arise once the problem is "solved." Fourth, firmly entrenched cultural values may strongly influence and delimit the range of policy options that can be brought to bear on a social problem, thereby resulting in a less-than-optimal "satisficing" strategy that appeases stakeholders but fails to address root causes. Fifth, predictions themselves have consequences and can change the dynamics of a situation being addressed by political action. Although conceived in the early part of the twentieth century, these explanations for unintended consequences have remained relevant and valid across a variety of contexts and issues.

For years, policy makers seeking to preserve the status quo have invoked the threat of unintended effects as a rhetorical strategy to prevent change. Political systems can become paralyzed and inert if the fear of unknown, unintended consequences exceeds the hope riding on some policy reform. Yet this is not to imply that the invocation of unintended consequences is a tool only of conserva-

tives, wielded against reform-minded liberals. For example, a recent review of climate change coverage in the United States and
Europe found that, in the United States, Democrats tend to use a
"Pandora's Box of catastrophe metaphor, whilst Republicans have
tended to emphasize the money frame and the scientific uncertainty frame" (Anderson, 2009, p. 174). In other words, liberals and
conservatives both fuel and capitalize on public skepticism and
fears by speculating about the unintended consequences of existing and proposed policies. The only difference is the particular nature of frightening frames employed on each side of the debate.

Sociologist Joel Best (2001) has described this phenomenon as
a "paradox of paranoia," which involves a belief that social destruction will result "...from unintended, ironic, potentially cata-
strophic consequences of social progress" (p. 7). While vast
segments of the public have faith in technology, many simultaneously believe that the byproducts of improved technology ultimately will destroy society through nuclear war, nuclear winter, global
warming, water and air pollution, resource depletion, or genetic
engineering. It is precisely this underlying cultural ethos that
makes invocations of unintended consequences so effective as a
rhetorical strategy in policy disputes. The fear of the unknown and
the potential for mass scale adverse consequences can constitute
compelling rationales for mobilizing constituencies and defeating
proposals for change.

Unintended Consequences, Issue Framing and Claims-Making

As a number of contemporary sociologists have observed, social
problems are products of collective definition rather than an objective assessment of science or social systems. Herbert Blumer is often credited for introducing the constructionist's perspective in his
articulation of social problems as collective behavior (Blumer,
1971). In his essay, Blumer posits that "a social problem exists
primarily in terms of how it is defined and conceived by society,"
(p. 300). In this view, problems are not social problems until the

public defines them as such; social problems do not "exist," they are "constructed."

The public arena refers to the discursive environment in which claims-making occurs (Hilgartner & Bosk, 1988). Broadly defined, social problems are "projections of collective sentiments rather than simple mirrors of objective conditions in society" (Hilgartner & Bosk, 1988, pp. 53–54). From this perspective, competition is a critical component directing how social problems are defined, framed, and packaged to advance a particular problem into the realm of public discourse. Human attention spans, media agendas and legislative dockets all have finite capacities, all too limited to respond to the multitude of problems that deserve public and private attention. Hence competition is keen among issue managers to focus attention on certain problems rather than others and to win widespread adoption of certain frames or ways of thinking about those problems and their solutions.

This competition of claims is evident in the construction and coverage of the biofuel debate in the United States, an on-going saga of science, geopolitics and individual behavior. The competition of claims surrounding the use of biofuels, however, is not a new issue in the American public domain. The issue has moved in and out of the public consciousness for more than forty years. Moreover, the use of unintended consequences in the construction of messages related to the use of fossil fuels is what brought biofuels as an alternative into the public agenda in the first place. This chapter analyzes this debate through a social constructionist examination of claims-making efforts, with a particular focus on claims about unintended consequences.[1]

---

[1] Examples are taken from *The New York Times* and *The Washington Post* archives as well as stories appearing on the AP wire service. These archives were retrieved through keyword searches including: "biofuels, biofuel debate, ethanol, ethanol debate, unintended consequences, Energy Act, gasohol, Surface Transportation Assistance Act." Supporting materials were retrieved through Internet searches and stakeholder blogs and websites.

Framing Biofuel

Liquid motor fuels such as biodiesel and bio-ethanol appear to be appealing alternatives to fossil fuels because they are produced from agricultural raw materials and can be mixed with normal petrol and diesel, with no engine adaptation required (Soetaert & Vandamme, 2009). Biofuels were first introduced to the American public as an alternative to other social problems arising from foreign dependence on oil and destruction of natural resources extracting fossil fuels. Yet, few policy issues are as complex as the current biofuel debate, a topic that triggers sharply polarized views.

Biofuels come in different forms and are made from different ingredients. The two most common biofuels are ethanol and biodiesel. Biodiesel is produced from fat or vegetable oil (National Renewable Energy Laboratory, 2010); in the U.S., the raw materials come from soybeans. Ethanol is produced from starchy sources such as corn or sugar cane. Ethanol has been at the crux of the U.S. biofuel debate. In common use, ethanol refers to a blend of biomass and petroleum fuel, meaning that gasoline is still a key (and majority) ingredient in the fuel. The paradox, therefore, is that investment in ethanol does not eliminate dependency on petroleum but arguably perpetuates it.

The Promise of Gasahol

Biofuels are not new potions but rather types of fuel that can be traced to the earliest days of the automotive industry. The first engines produced in Detroit ran on an alcohol-based form of corn ethanol, otherwise known as moonshine. These engines powered Ford vehicles for 13 years. Prohibition eventually mandated that alcohol-based fuels be replaced with petroleum, a development that has led to questions and conspiracy theories about the role of John D. Rockefeller and Standard Oil Company in the funding and organization of the Prohibition Movement (e.g., Tickel, 2009). It is rumored that Rockefeller donated over $350,000 to the Anti-Saloon League.[2]

The term "biofuel" was coined in the 1970s in part by President Jimmy Carter, who introduced biofuels as part of an energy plan to reduce reliance on oil in general and spite the Soviet Union in particular. More specifically, he aimed to decrease gas consumption, in part by promoting a new bio-fuel, a 10-percent ethanol gasoline blend called gasahol. The move toward an ethanol-based fuel had several political motives. First, the administration could punish the Soviet Union by withholding grain, while keeping domestic grain prices high. Second, the higher grain prices would benefit the farm belt in an election year. Third, ethanol-based fuel would please the driving public by increasing supplies of automotive fuel while cutting the reliance on foreign oil. The reduced reliance on foreign oil would in turn strengthen the U.S. dollar (Lyons, 1980). The public was supportive of the idea; however, the United States simply did not have the capacity to grow enough corn or the technology to efficiently convert corn to fuel. Despite the benefits, Carter was unable to deliver on the promise to mass produce gasahol.

Gasahol received high levels of political support in the 1980s primarily from farm-state political leaders, who could win favor with their constituents by advocating for a new market in corn-based-ethanol. Proponents included agri-business giants who were heavily invested in the corn industry as well as the farmers and farming communities that yielded corn crops. Even among supporters, however, there were murmurs of unintended consequences of the production of ethanol-based fuel. One Wisconsin-based representative warned that increased ethanol production needed to be coupled with gasoline rationing programs. His opinion was shaped by the potential energy costs for storing grain which would offset any energy saved by producing and burning ethanol compared to fossil fuels (Lyons, 1980).

Opponents of ethanol fuels immediately recognized and voiced claims about the unintended effects of gasahol. For example, the

---

[2] http://www.spartacus.schoolnet.co.uk/USAprohibition.htm

American Petroleum Institute warned that ethanol blends would cause high costs, bad starts, low mileage, bad driveability and engine damage. The oil industry also complained that ethanol was impractical for nationwide distribution because it could not be transported by pipeline. Despite claims about ethanol-based biofuels, ethanol blenders and producers received large scale federal incentives to develop their biofuels.

The Energy Policy Act of 1978 was the first step toward building the ethanol industry. Congress passed a series of bills providing protective tariffs as well as investment credits and loan guarantees for new ethanol plants (Tyner, 2008). Another alcohol fuel, methanol, had also been promoted as a gasoline alternative; however, methanol received no tax credits or protective tariffs. Initially, federal biofuel policies were developed to help launch the biofuel industry during its infancy. Federal policy played a key role in helping to close the price gap between biofuels and cheaper petroleum fuels.

Through the 1980s and 1990s the United States continued to offer tax exemptions for ethanol-based fuel while increasing gas taxes (e.g., 1982 Surface Transportation Assistance Act). In 1988, the Alternative Motor Fuels Act funded research and offered fuel economy credits further incentivizing research and development for ethanol-based fuels. Exemptions and subsidies continued in the United States through the 1990s, dropping slightly at times. Ethanol production was further supported by a strong coalition of farm-state legislators, grain processors and trade groups who were able to shepherd policy initiatives through Congress through the 1980s (Weiss, 1990). Most of the lucrative subsidies ended up benefiting one particularly large ethanol producer during this time. Political lines were blurred because the ethanol industry had strong Washington connections on both sides of the aisle, Bob Dole and Tom Daschle among the most avid supporters (Weiss, 1990).

Mass production of ethanol, however, had hidden costs. There were signs of these unintended effects in the first decade of the ethanol industry. For example, a 1986 report issued by the U.S.D.A.'s Office of Energy contended that the subsidies would drive up the demand for corn and increase consumer food costs. Additionally, the report claimed that any benefits to corn growers would come at the expense of soybean farmers. A domino effect would then occur as soybean prices dropped because livestock feed, a byproduct of ethanol production, competes with soybean meal. Members of the influential lobbing group became so upset with the report that the U.S.D.A director received considerable political backlash threatening his job (Weiss, 1990).

The political momentum for ethanol began to fade in the early 1990s. At this time the National Corn Growers Association (NCGA) realized the need for a popular appeal to the masses. The framing of biofuel as an alternative energy started to take shape at the annual conference in Kansas, NCGA spokesperson Mike Bryan explained in 1990:

> We can no longer focus on selling the fact that it's a farm product and what it's going to do for the farmer....We have to focus on clean air. We also have to focus on the reduction in oil imports. We realize that in Manhattan, most people think corn comes from a grocery store....If we tell them we're helping to clean their air, that's something they can relate to.

The Twenty-First Century Biofuel

The twenty-first century marked an increased level of public concern associated with climate change and geo-politics. Fuel formed in the earth from plant or animal remains (i.e., fossil fuels like petroleum, natural gas, and coal) continued to grow in demand and diminish in supply. Concerns over foreign dependency on oil, concerns over climate change associated with fossil fuels, and concerns over the depleting reserves of fossil fuels added to the practical and emotional urgency for alternative fuels (Sagar & Kartha, 2007). Proponents of ethanol claimed that biofuels could decrease greenhouse gases while providing green jobs for American workers. All the while, the ethanol industry continued to receive subsidies and incentives to establish technology that could meet high market demands.

Beginning in 2004, crude oil prices continued a steep upward climb. This rapid increase in the crude oil price in combination with the ethanol subsidy led to a tremendous boom in construction of ethanol plants. The real "ethanol boom" began in 2005 with the Energy Policy Act. By this time, the ethanol industry had been swelling under the surface for decades. The agri-business was perfectly situated to accommodate mass production but needed the higher demand. Prominent, wealthy business leaders invested in ethanol and expected a future market to reap the rewards of their investments. The United States seemingly believed in its capabilities to harvest massive amounts of corn and profit from it. And, once again, a presidential election was on the horizon. Corn ethanol offered candidates a platform to appeal to Midwestern corn-producing states, which also happened to include several key swing states (e.g., Idaho and Michigan). The combination of increased public concern toward the environment, the increased capacity for production (compared to Carter's situation three decades earlier) and the potential profit for stakeholders and struggling farm states created the perfect storm for the ethanol industry.

However, not everyone would benefit from an ethanol takeover. Opponents of ethanol-based biofuels included a number of fiscally conservative citizen groups, objecting to ethanol mandates on several terms including the loss of a free marketplace. Opponents argued that an ethanol mandate would only benefit Midwestern corn farmers and ethanol producers. For everyone else, mandates for biofuels would simply raise the price of gas. The higher the level of production would drive up corn prices and make it more difficult for ethanol producers to meet demand. In a 2005 letter to Congress, the Heritage Foundation outlined a number of related unanticipated economic outcomes, such as higher gas prices and higher food prices. The group claimed that: "the ethanol mandate is an anti-consumer provision. It benefits special interests at the expense of the driving public. As such, it has no place in an energy bill that seeks to make energy more affordable for the American people" (Lieberman, 2005).

Warnings of high gas and food prices seemingly fell on deaf ears. The boom that followed included President George W. Bush's renewable energy fuel standard starting at 4 billion gallons in 2006; this legislation essentially mandated the fuel industry to acquire a certain percentage of its fuel from renewable sources. The 2007 Energy Independence and Security Act (EISA) boosted this standard even higher, calling for 35 billion gallons of biofuel by 2022.

When President George W. Bush signed the EISA in December of 2007, he announced "a major step forward in expanding the production of renewable fuels, reducing our dependence on oil, and confronting global climate change." Bush explained that the mandates will "increase our energy security, expand the production of renewable fuels, and make America stronger, safer, and cleaner for future generations" (Energy Independence and Security Act of 2007 (H.R. 6), 2007).

Opposition to ethanol-based biofuel became more apparent following the Renewable Energy Act. Economists and environmentalists questioned why the standards outlined in the energy act favored ethanol compared to other types of renewable fuels. Economists questioned the rationale for the long term subsidies enjoyed by the ethanol industry. Environmentalists questioned the true energy savings for ethanol. Opponents of the subsidies argued that the ethanol industry had outgrown the need for the federal incentives and protections designed in the 1970s; referring to the infant ethanol industry, one columnist noted that "the infant has grown into a strapping behemoth with a powerful sense of entitlement and an insatiable appetite for ethanol's primary feedstock: corn" (Runge, 2010, p. 10).

The public began to pay attention to the subsidies for ethanol, which had seemingly escaped serious public scrutiny for decades. Criticisms grew sharper, demarcating a new debate surrounding the economics of ethanol. Ethanol proponents, such as the American Coalition for Ethanol, countered these claims by reminding the public of the looming climate crisis and vilifying ethanol opponents as special interest oil lobbyists.

...asking the oil industry to sell ethanol is like asking cattlemen to sell tofu—it's not their product and they'd rather not use it. The transportation fuel supply is a pretty profitable status quo for the petroleum industry, and understandably, they have no reason to want things to change. But as Americans, we do want things to change and understand that things must change. The threat of global warming and the reality of a limited oil supply are two red flags on the horizon, reminding us that clean, renewable alternatives are needed. (Jennings, 2007, p. 1)

In 2008, the entire premise of biofuels came under attack following two *Science* magazine articles claiming that ethanol could create more problems than solutions. The magazine published two studies that found that almost all biofuels caused more greenhouse gas emissions than conventional fuels if the full environmental costs of production are considered (see, Searchinger et al., 2008; Fargione, 2008). Immediately, 600 newspapers picked up the story (Tickel, 2009). The headlines focused on the unintended effects of biofuels, including: deforestation, world hunger and global warming. Biofuels suddenly turned from the clean energy solution to the clean energy scam (Grunwald, 2008).

The publication of the *Science* magazine articles marked a dramatic shift in the debate about biofuels. Initially biofuels had been framed as a solution to problems of global warming; proponents had promised that ethanol would reduce greenhouse gas emissions and curb the global demands for carbon-based petroleum product. Following the *Science* magazine articles, several controversies surfaced over the unintended social and environmental impacts of increased corn ethanol production and use. The tide had turned on biofuel, as opponents now spanned the communities of science, international human rights, and environmental policy.

Anti-Ethanol Frames: Displaced Ecosystems & Economies

Opponents of biofuels argue that the higher market demand for corn in combination with incentives to grow corn creates a natural imbalance for both the economy and numerous ecosystems threatened by the spread of corn crops. This type of unintended consequence illustrates a ripple effect attributable to the increased demand for corn. Corn crops, like other food crops, require land in order to harvest enough to meet demands created by mandates and fuel standards requiring certain percentage of ethanol-based fuels. Because land is a finite resource, other crops are displaced in order to accommodate corn crops. Similarly, forests and wetlands are cleared to accommodate corn and other displaced crops. The higher value of corn creates a ripple effect throughout the global agricultural infrastructure; corn displaces other crops and infringes on natural ecosystems. This, in turn, increases the level of greenhouse gases as large parcels of land are cleared in order to support the growth of corn crops. Some of those parcels of land currently support thick forests that balance greenhouse gas emissions, protecting the ozone from harmful chemicals (Ehrlich & Ehrlich, 1981). Thus, clearing a portion of carbon-capturing forest in order to plant corn for ethanol arguably has the opposite effect as originally intended for alternative fuels. Often, claims-makers use the example of the destruction of Indonesia and Malaysia wetlands to illustrate the potential consequences of biofuels. In Indonesia and Malaysia, areas were cleared to harvest palm oil in response to European demands for biofuels. Heavily forested peat swamplands were logged and drained in order to grow oil palms, the fruit of which was used as the source of the oil. The process also increases the likelihood of wildfires further diminishing forests and wetlands.

Another scenario involving the unintended consequence of land use to support corn crops is the harmful impact of invasive species. Biofuel crops will likely be cultivated on lands surrounded by sensitive forest, prairie, desert, and riparian areas as well as by rangelands and agricultural commodities (DiTomaso, Barney, & Fox, 2007). The sensitivity of these ecosystems leaves them vulnerable to invasive corn crops. Similarly, as corn moves in, other essential food crops are displaced. The movement of displaced crops to peripheral areas of fields and grasslands can result in the spread of non-native plants spreading into natural habitats, thereby pushing out naturally occurring plant species. People, too, can be displaced by heavily subsidized crops. In Colombia, where acreage dedicated to the production of palm oil has more than doubled in the past four years, armed groups are driving peasants off their lands to create palm-oil plantations.

Starved Soil and Starving People

If forests are left intact in order to neutralize greenhouse gas emissions, other crops will need to be displaced in order to facilitate corn growing. Some agricultural groups claim this is problematic for the environment as well. First, opponents point out that corn is a particularly damaging crop because farmers tend to use high levels of pesticides that disperse in the air, contaminate the soil and leak into water supplies. Second, farmers who previously rotated corn crops with soybean crops are rewarded for growing only corn. When crops aren't rotated the soil becomes barren, losing its nutrients and causing erosion requiring the farmers to use even more fertilizers and pesticides, which eventually end up in the water supply. Similarly, the United Nations Energy Report warned that U.S. policy favors large-scale production and consequently the use of mono-cropping, which could lead to biodiversity loss, soil erosion and nutrient leaching (Bringezu et al., 2007). The report also explained that a rapid increase in liquid biofuel production will place significant stress on land and water resources at a time when global demand for food and forest products is already growing rapidly (Bringezu et al., 2007). This highlights a second-order unintended consequence of biofuel production, namely displacing food crops for fuel or using corn for fuel instead of food.

This chain of events is referred to by claims-makers as "remorseless economics." The World Bank claims that biofuels will result in world hunger due to rising food prices. Similarly, *The Economist* (2007) called the rise of biofuels "the end of cheap food." Lester Brown (2009) of the Earth Institute, points out that not every kernel of corn diverted to fuel will be replaced. Diversions raise food prices, so the poor will eat less (Brown, 2009). Claims-makers commonly use the corn riots in Mexico to illustrate how dire the food shortage has become. For example, *The Christian*

Science Monitor (Nothstine, 2007) highlighted the incident explaining:

> In Mexico, where corn is a staple, rapidly rising prices for tortillas sparked open revolt. Tortilla prices skyrocketed more than threefold last year. Protesters took to the streets in Mexico City, compelling the normally free market-minded President Felipe Calderón to cap prices at 78 cents per kilogram.

Proponents of Ethanol: Reframing the Biofuel Debate

In contrast to opponents' frames of displacement and depletion, proponents of biofuels have attempted to reframe the biofuel debate, focusing on the need to reduce dependency on foreign oil and concomitantly increase homeland security. Yet since 2008 they have spent much of their time responding to claims of imminent environmental and social disasters rather than proactively constructing their own claims about the benefits of ethanol. For example, the American Coalition of Ethanol responded to attacks on biofuels by claiming ethanol might not be the perfect solution but is the only alternative to fossil fuels.

> It's disappointing that some in the environmental community continue to have an irrational and unsophisticated notion of how to reduce fossil fuel use...Some say the solution is to get rid of corn-based ethanol today, in hopes that some other potentially promising, but not yet commercialized fuel will be available tomorrow. The result would be more pain at the pump and more pollution for the planet. Ethanol is the only commercially available alternative to gasoline today, and removing it from our nation's fuel supply would mean more oil use—and we ought to learn from the painful and ongoing lesson in the Gulf of Mexico that more oil is simply not a sustainable path. (Jennings, 2010, p. 1)

Among the proponents of ethanol biofuels are prominent investors such as General Electric, British Petroleum, Ford, Shell, Cargill and the Carlyle Group (Grunwald, 2008). With such heavyweight supporters, the ethanol industry is far from giving up the fight in the face of claims about the harmful unintended consequences of ethanol production and use. Shifting their focus away from decreased greenhouse gases, proponents of ethanol have adopted an

"All American" frame. This frame hinges on American values and patriotism, thereby attempting to move the debate away from unintended consequences.

Proponents claim that ethanol production will create American jobs and support American farmers. Appeals based on patriotic calls for American innovation and leadership tend to tap into concerns about energy independence. These appeals build support for green energy policies because they draw on widely held shared values. This frame reprises the early frames presented in the Carter Administration era. A spokesperson for the Renewable Fuels Association sums up a worst-case scenario for Americans if ethanol subsidies are recanted:

> Eliminating investment in ethanol and other renewable fuels will result in reducing domestic ethanol production, increase unemployment, and increase America's dependence on imported oil from OPEC members like Saudi Arabia and Venezuela whose policies are often at odds with ours and ethanol from Brazil, where rainforest destruction happens with regularity. (Dinneen, 2010. p. 1)

This frame emphasizes the patriotism behind supporting ethanol production and usage: ethanol is pro-American. Pro-ethanol framing often embraces widely shared American values including the free market system and an individual's freedom to choose which type of fuel suits him/her. The American Coalition of Ethanol consists of industry stakeholders including Corn Growers Association, several energy cooperatives, farmers and agri-business executives. Lars Herseth (2010), the group's president, summarized this sentiment in an editorial letter entitled, "American Consumers Deserve Fuel Choice."

> Today, petroleum has a de facto mandate at American gas stations. Instead of the oil industry or the government making the decision about what fuel to put in your vehicle, the consumer should have the right to choose. The longer we wait to provide Americans with real fuel choice, the more consumers will pay at the pump, the more the environment will suffer, and the more our nation will rely on foreign sources of oil. (p. 1)

Morality also comes into play as a shared value and counter claim among claims-makers supporting ethanol biofuels. For example, in response to a proposal that would allow states to opt out of renewable fuel standards, the American Agriculture Movement adopted a dramatic frame linking oil dependence and the death of American soldiers.

> I would call the proposal anti-ethanol, anti-American and downright treasonous....The death of each American soldier killed with weapons bought or produced with the oil money that America sends to the Middle East would be on his head and clearly he is aiding and financing our enemy. The very act of introducing this legislation and attempting to get such a bill passed is in my opinion clearly treasonous. (Matlack, 2010, p. 9)

Discussion and Conclusion

The importance of studying unintended consequences of public policy options and decisions is a focal point of "Weathercocks and Signposts," through its assertion that social marketing is inherently ill-suited to serve as a mechanism for meaningful and broad-scale social change. The notion of unintended consequences similarly serves as a useful framework for examining the manifest and latent impacts of various environmentally related technologies, such as ethanol.

The most politically significant articulation of the unintended consequences of ethanol occurred with the publication of articles in *Science* magazine in 2008. It is difficult to overstate the impact of these articles on the ethanol debate. Although they repeated several claims that had been made in the past, they legitimized and publicized the potential unintended consequences of ethanol to an unprecedented degree. In so doing, they also illustrated the process of what sociologist Stephen Hilgartner (1990) describes as the popularization of science. Hilgartner asserts that the culturally dominant view of science is a function of a two-staged model: first, scientists develop what he labels "genuine knowledge"; second, popularizers distribute streamlined versions to the public. Hilgartner is highly critical of the reinterpretation and distribution of science through popular media: "at worst, it constitutes a form of 'pollution'—the distortion of science by outsiders" (p. 519).

Science-based issues such as alternative energy are particularly vulnerable to oversimplified frames of unintended consequences, as they are often ambiguously explained by scientists (Zehr, 1994) and misunderstood by reporters. This vacuum of explanation and understanding creates an opportunity for interpretive claims-makers to interpret and reframe issues, while also creating the conditions for high levels of interpretive flexibility on the part of the public (e.g., Silverstone, 1991; Wynne, 1991).

In the case of biofuels, the specter of potential unintended consequences has played a significant role in policy debates since the 1970s. Claims-making proponents initially touted biofuels as a means of addressing the unintended consequences of America's addiction to oil, only to see claims-making opponents criticize biofuels on the basis of a different, and potentially more serious, set of unanticipated consequences.

Yet common to all types of unintended effects is the difficulty in assessing their validity. Unintended consequences usually cannot be measured or studied at the time at which the claim of their inevitability is made. Unintended consequences usually are the result of a complex chain of micro- and macro-level processes that do not lend themselves to clear-cut analysis or unambiguous causal linkages. Consequences alleged to result from one policy action may in fact result from another or from entirely separate chain of events or set of incidents. For example, opponents of ethanol claim that:

> Deforestation accounts for 20% of all current carbon emissions. So unless the world can eliminate emissions from all other sources—cars, power plants, factories, even flatulent cows—it needs to reduce deforestation or risk an environmental catastrophe. (Grunwald, 2008)

To this claim, proponents ask, "what percent of deforestation is a result of 'indirect land use' or removal of forests for ethanol producing crops?" Experts largely agree that we do not have the scientific ability at present to measure the indirect land use change effects (Dale, Efroymson, & Kline, 2011). Methods to assess indirect land-use emissions are controversial. All quantitative analyses to date have either ignored indirect emissions altogether, considered those associated from crop displacement from a limited area, confused indirect emissions with direct or general land-use emissions, or developed estimates based on a static framework of today's economy (Melillo et al., 2009).

A second common characteristic of unintended effects is their reliance on fear appeals: higher gas prices, global warming, diminished national security and starving children are only a few of the fearful frames that have been used in this debate. Fear appeals are particularly effective for framing issues related to science because the "science" behind the claims is often misunderstood by policy makers and citizens. These characteristics—i.e., fear appeals and an inability to substantiate or refute the validity of presumed and alleged unintended consequences—have their own dire implications for public understanding of biofuels as a social problem. While the public remains mired in the debate about whether promotion of ethanol indirectly kills people, global temperatures rise, children continue to die of malnutrition, and the world's addiction to oil continues unabated. In the words of Suleiman J. Al-Herbish, a minister for OFID, a sister organization to OPEC:

> ...There is a risk that we might spend the next 30 years debating the merits of feeding cereals to cars. This time the situation though is different, as the entire world's population will be affected if we fail to deal with the challenges of climate change mitigation, providing clean energy and ensuring food security, all of which are interrelated and need to be tackled together. (*OPEC Bulletin*, 2009, p. 77)

References

Al-Herbish, S.J. (2009). *OPEC Bulletin.* Vienna, Austria: Organization of the Petroleum Exporting Countries.

Anderson, A. (2009). Media, politics and climate change: Towards a new research agenda. *Sociology Compass, 3*(2), 166–182.

Best, J. (2001). Social progress and social problems: Toward a sociology of gloom. *The Sociological Quarterly, 42*(1), 1–12.

Blumer, H. (1971). Social problems as collective behavior. *Social Problems, 18*(3), 298–306).

Brown, L. (2009). *Plan B 4.0: Mobilizing to save civilization.* New York: W.W. Norton & Company.

Bringezu, S., Schütz, H., O'Brien, M., Kauppi, L., Howarth, R. W., & McNeely, J. (2007). UN energy report: Assessing biofuels. Geneva: United Nations Report.

Chappell, T. (1994). *The soul of a business: Managing for profit and the common good.* New York: Bantam Books.

Cho, H., & Salmon, C. T (2007). Unintended effects of health communication campaigns. *Journal of Communication,* 57, 293–317.

Dale, V., Efroymson, R., & Kline, K. (2011). The land use–climate change–energy nexus. *Landscape Ecology,* 26, 755–773.

Dinneen, B. (2010). Ethanol industry counters environmental stall tactics. Retrieved from http://www.ethanol.org/pdf/contentmgmt/Ethanol_industry_counters_environmental_stall_tactics_6152010.pdf

DiTomaso, J. M., Barney, J. N., & Fox, A. M. (2007). *Biofuel feedstocks: The risk of future invasions. Science Source for Food, Agriculture, and Environmental Stall Tactics.* Retrieved from http://www.ethanol.org /pdf/ contentmgmt/Ethanol_industry_counters_environmental_stall_tactics_6152010.pdf

Ehrlich, P. R., & Ehrlich, A. H. (1981). *Extinction: The causes and consequences of the disappearance of species.* New York: Random House.

Energy Independence and Security Act of 2007 (H.R. 6), 2007

Fargione, J., Hill, J., Tilman, D., Polasky, S., & Hawthorne, P. (2008). Land clearing and the biofuel carbon debt. *Science, 319,* 1235-1238.

Grunwald, M. (2008, Mar. 27). The Clean Energy Scam. *Time Magazine.* Retrieved from http://www.time.com/time/magazine/article /0,9171,1725975,00 .html

Herseth, L. (2010). American consumers deserve fuel choice. Retrieved from http://www.ethanol.org/pdf/contentmgmt/Herseth_fuel_choice_oped_April_2010.pdf

Hilgartner, S. (1990). The dominant view of popularization: Conceptual problems, political uses. *Social Studies of Science,* 20, 519–539.

Hilgartner, S., & Bosk, C. L. (1988). The rise and fall of social problems: A public arenas model. *The American Journal of Sociology, 94*(1), 53–78.

Jennings, B. (2007). Ethanol: Breaking the crude oil mandate. Retrieved from http://www.ethanol.org/pdf/contentmgmt/Breaking_the_Oil_Mandate_oped_s pring_07.pdf

Jennings, B. (2010). Ethanol industry counters environmental stall tactics. Retrieved from http://www.ethanolrfa.org/news/entry/ethanol-industry-counters-environmental-stall-tactics/

Lieberman, B. (2005). Keep ethanol out of the energy bill. Heritage Foundation Report. http://www.heritage.org/research/reports/2005/04/keep-ethanol-out-of-the-energy-bill

Lyons, R. D. (Jan 8, 1980). U.S. Struggles for Gasohol Plan. *New York Times*. Retrieved from http://proquest.umi.com.proxy2.cl.msu.edu/ pqdweb?index= 6&did=111760398&SrchMode=1&sid=2&Fmt=10&VInst=PROD&VType=PQ D&RQT=309&VName=HNP&TS=1292698623&clientId=3552

Matlack, L. (2011). AAM urges action against Inhofe anti-ethanol proposal. Washington: American Agriculture Movement Report. Retrieved from http://www.aaminc.org/pressreleases/

Melillo, J.M., Reilly, J.M., Kicklighter, D.W., Gurgel, A.C., Cronin, T.W., Paltsev, S., et al. (2009). Indirect emissions from biofuels: How important? *Science*, 326, 1397–1399.

Merton, R. K. (1936). The unanticipated consequences of purposive social action. *American Sociological Review, 1*(6), 894–904.

National Renewable Energy Laboratory. (February 2, 2010). *Biofuels*. Retrieved January 10, 2011, from http://www.nrel.gov/learning/re_biofuels.html

Nothstine, R. (July 27, 2007). The unintended consequences of the ethanol quick fix. *Christian Science Monitor*. Retrieved from http://www.csmonitor.com /2007/0727/p09s02-coop.html

Runge, F. (2010). Biofuel backlash: Subsidies for corn ethanol are hurting people and the planet. *Technology Review*. Retrieved from http://www.technology review.com/energy/25109

Sagar, A.D., & Kartha, S. (2007). Bioenergy and sustainable development? *Annual Review of Environment and Resources*, 32, 131–167.

Salmon, C.T. (1992). Bridging theory 'of' with theory 'for' communication campaigns: An essay on ideology and public policy. In S. Deetz (Ed.), *Communication Yearbook 15* (pp. 346–358). Newbury Park, CA: Sage Publications.

Searchinger, T., Heimlich, R., Houghton, R. A., Dong, F., Elobeid, A., Fabiosa, J., et al. (2008). Use of U.S. croplands for biofuels increases greenhouse gases through emissions from land-use change. *Science*, 319, 1238–1240.

Silverstone, R. (1991). Communicating science to the public. *Science, Technology, & Human Values*, 16, 106–110.

Soetaert, W. & Vandamme, E. J. (2009) Biofuels in perspective, In Soetaert, W., & Vandamme, E.J. (Eds.) *Biofuels*. John Wiley & Sons, Ltd.: Chichester, UK. doi: 10.1002/9780470754108.ch1

Tickel, J. (2009). *Fuel: Change your fuel, change your world.* United States: Cinema Libre Studio.

Tyner, W. E. (2008). The US ethanol and biofuels boom: Its origins, current status, and future prospects. *Bioscience, 58,* 646–653.

Weiss, M.J. (April 1, 1990). The high-octane ethanol lobby. *The New York Times Magazine* Retrieved from http://find.galegroup.com.proxy2.cl.msu.edu /gtx/infomark.do?&contentSet=IAC-Documents&type=retrieve&tabID=T004 &prodId=STND&docId=A175483744&source=gale&srcprod=STND&userGro upName=msu_main&version=1.0

Wynne, B. (1991). Knowledges in context. *Science, Technology and Human Values* 16, 111–121.

Zehr, S. (1994). Method, scale, and socio-technical networks: problems of standardization in acid rain, ozone depletion, and global warming research. *Science Studies* 7, 47–58.

CHAPTER FOUR

Greenwashing to Green Advocacy: The Environmental Imperative in Organizational Rhetoric

Brant Short

Organizational rhetoric transcends advertising and public relations and includes strategic messages related to topics such as public policy, ethics, and cultural values. Many organizations recognize the need to protect the environment and have responded with green advocacy, promoting environmental awareness and suggesting significant changes in lifestyle. Critics view green advocacy as a form of greenwashing in which organizations proclaim a false environmental commitment in order to create positive perceptions by consumers. However, organizational rhetoric is ubiquitous in today's global mediated culture, and, if used appropriately, it offers a powerful and positive voice for promoting sustainable practices among multiple stakeholders. Citizens, scholars and activists need a different perspective, grounded in the traditions of rhetorical studies, to more fully identify, critique, and promote green advocacy by organizations. In this chapter, I examine the rhetorical strategies and tactics employed by environmental advocates. I then review the rise of organizational rhetoric and its recent move toward values and public affairs, including green branding and eco-marketing. I conclude by assessing the use of greenwashing as a viable construct and offer an alternative model for assessing green advocacy.

Environmental angst defines every aspect of life in the twenty-first century. From energy to education, from transportation to art, from our birth to our demise, people are asked to think every day about their role in saving, or destroying, our planet. In the second half of the twentieth century, many Americans accepted the need for significant changes in resource consumption and energy use and in turn urged others to join the environmental crusade. In the past decade, however, a sense of urgency for collective action has given way to complacency and critics of the environmental movement have gained both status and power. David Norman of the WWF-UK concludes that the global environmental movement is at a crossroads. It can continue to seek incremental changes through a market system of green consumerism, or it can "begin to inject new urgency into the environmental debate" and seek more radical changes in behavior (Crompton, 2008, p. 2). Creating a "new urgency" has both challenged and frustrated activists and policymakers. But there are signs that most people are ready to act, but they need both guidance and motivation. One potentially rich source of guidance and motivation lies in organizations and their public communication narratives.

Tensions in the Environmental Agenda

The 2008 report "Weathercocks" authored by Tom Crompton and distributed through the auspices of the World Wildlife Fund offers a detailed assessment of the environmental movement and its failures to ignite a revolution in values, attitudes and behaviors among the world's consumers. The report is well documented and challenges simplistic approaches to changing human patterns of consumption and materialism. Crompton has identified a set of at least four tensions that confront policy-makers and environmental advocates. Recognizing and resolving these tensions seem to be at the heart of the next series of actions by the leaders.

Small Steps Versus Radical Change. The failure of using targeted and specific changes that require little sacrifice seems apparent. But invoking radical change provokes defensiveness and gives environmental critics a platform to proclaim a false commitment to moderation. Can the environmental community find a middle road?

The Ends Versus Means Dilemma. This theme emerges throughout Crompton's report. Do we need to make sure people take actions for the right reasons or should we be satisfied with the outcome? If behaviors are changed that will protect the earth, should we care if the reason is selfish or altruistic?

Focus upon Products Versus Practices. Should the environmental community look toward solutions in technology and innovation, or should the solution be focused upon levels of consumption and lifestyle choices? Does one solution inherently diminish accomplishing the other?

Persuasion Versus Regulation. In Western cultures that revere human rights and freedom of choice, regulation is often viewed as a last resort for changing behavior. But if persuasion fails to invoke meaningful changes, should governments begin strict regulation of consumer practices?

These tensions challenge anyone thinking about strategies to foster meaningful and long-term changes in how people approach the environment. Like all tensions, people tend to gravitate toward one side or the other, but as Crompton has shown, innovative and practical approaches are needed, and in some cases the tensions may need to be addressed in unconventional, even radical ways.

Organizational Advocacy and Greenwashing

Organizations hold a privileged place in contemporary society. They typically are not part of the established political order but hold great power in shaping how people construct meaning and respond to the world. Moreover organizations define nearly every part of our daily life and include not only businesses but also universities and schools, hospitals, nonprofit agencies, social groups, and religious institutions, among many others. They are inherent parts of our life and fulfill needs that cannot easily be accomplished individually. Even though humans have turned to organizations for thousands of years, in the twentieth century organizations emerged to define every sector of our life. Robert Heath (2006) contends that humans, as individuals and in groups, confront a world of chaos, entropy and turbulence. In turn, we create organizations to help us manage these risks, reduce uncertainty, and help us make enlightened choices. And it is through organizational advocacy that shared meanings are created, goals are negotiated, conflicts are managed, and risks are contained. Heath writes: "Organizations play a substantial role in creating ideas–the shared sense of social reality that serves the making of choices, individual and collective" (2006, p. 95). This may be more profound than it might seem. The creation of knowledge is almost always founded in an organization, shared by an organization, and managed by an organization. With such a privileged and significant position in the cultural landscape, organizations often employ advocacy to shape behavior and attitudes among stakeholders.

We receive thousands of messages daily that are prepared and distributed by organizations. These are not random; they are strategic messages designed to inform, persuade, reinforce or create shared meaning with organizational stakeholders. Heath (2006) writes that organizations use public discourse for multiple reasons, including:

> ...to cocreate shared social meaning, negotiate relationships, influence and yield to influence, create and resolve conflict, distribute resources, manage power resources, exert and yield to control, manage risks, shape and respond to preferences, work to resolve uncertainty, foster trust, engage in support and opposition, distribute rewards and costs, foster interdependency, and make enlightened choices. (p. 98)

Organizations succeed or fail in a given time and place, and managing perceptions among their stakeholders is essential to such success or failure. Advertising and public relations are obvious examples of organizational advocacy, but it also includes speeches, interviews, websites, governmental testimony, corporate reports, newsletters and other texts crafted for public consumption. Moreover, citizens demand that organizational leaders interact with their constituencies, especially in times of crisis. Consumers see the organization as a concrete group of people, with a specific leader, and demand a human face in these situations. During product recalls, environmental disasters, and ethics scandals, organizations have made leaders available in many different public forums to answer questions and represent their organization as a single entity.

In the last twenty years many organizations have turned to environmental messages in presenting their organizational identity to selected audiences. These green messages include advertising and public relations as well as statements of organizational values and goals. What has prompted this shift away from simply advertising products and services? Why has green branding grown so dramatically in the last 20 years? There seem to be at least five goals prompting green advocacy: to get new customers; to share organizational values; to create or to change an organization's image; to improve interactions with other organizations; and to foster a positive culture for employees. Examples include advertisements that tout natural and recycled product components, the environmental impact of a given product or service, reductions in energy used to design, manufacture and/or distribute a specific product, and organizational commitments to protect the environment in tangible and/or long-term ways.

The rise of green advocacy prompted concerns of greenwashing by organizational critics. What is greenwashing and how has it been used? The term emerged in the 1970s as a label to identify unethical actions by businesses in regard to environmental issues. There are many published definitions of greenwashing and all refer to unethical conduct by corporations. A typical definition appeared in *Natural Life Magazine*: "Greenwash—verb: the act of misleading consumers regarding the environmental practices of a company or the environmental benefits of a product or service" (Priesnitz, 2008, p. 14). The term stems from the older term "whitewashing," which suggests a concealment of faults or mistakes. Typically greenwashing is used to label any organizational message that may be misleading, including exaggeration, omission, and even falsification. One scholar notes that most corporations do not use outright lying, "Rather, they bend the truth or misrepresent their ecological stances. The deception often lies in the emphasis corporations place on their ecological projects, rather than in the existence of the projects themselves" (Voss, 2009, p. 674). To combat greenwashing in corporate discourse, some scholars suggest that increased regulation and/or legislation is the answer. Gibson (2009), Voss (2009), Laufer (2003), and Munshi and Kurian (2005) all call for giving governmental agencies increased power to regulate corporate discourse and punish organizations that practice greenwashing.

A notable study of greenwashing, often cited in popular and scholarly accounts (Priesnitz, 2008; Gibson, 2009), comes from TerraChoice, a "science based marketing firm" located in Canada. The firm examined common household products making green claims in big box stores in Canada and the United States. They found "2219 products making 4996 green claims, with over 98% of retailers committing at least one of the Seven Sins of Greenwashing" (Mulch, 2009). The list of seven sins of greenwashing reflects various ways consumers might be misled by an organization. It can be outright falsification or it might be vagueness or not listing all environmental tradeoffs. The greatest sin identified was hidden trade-offs at 70% (using a narrow set of attributes to claim being green), and the least reported sin was fibbing (false claims) at 1%. Other categories include: the sin of irrelevance (7%), the sin of no proof (60%), the sin of vagueness (51%), the sin of worshiping false labels (23%), and the sin of the lesser of two evils (5%) (Mulch, 2009).

Greenwashing has entered the popular lexicon because it allows critics to easily label unethical efforts by corporations, and it affirms our suspicion of any business claiming to do good. The term is colorful, memorable, and provides an indictment without a detailed assessment of a specific claim. But its wholesale application to all forms of green advocacy calls into question its ultimate utility. If nearly all organizational efforts to illustrate environmental values of products and services commit one of the sins of greenwashing, as the TerraChoice study indicates, then consumers have little reason to search for and purchase green products, creating a market disincentive to embrace sustainable practices by organizations. I see three problems in the current use of greenwashing as a way to describe organizational advocacy.

First, greenwashing is an ideological term and not a descriptive/analytical term. When used for political purposes and to persuade people to take action, the use of greenwashing makes sense. But scholars, journalists and consumer advocates sometimes use the term as a description of a type of environmental message, without precision or clarity in application.

Second, greenwashing is an absolute concept and does not allow for judgments in degree and scope of the action. In its usage the term paints products, services and/or organizations in a binary judgment system. There is no place for discussions of messages that are greater or lesser on the scale of attempting to shape opinion in an ethical (or unethical) manner.

Third, greenwashing is used to label deception, omission or lack of clarity. The application of greenwashing may be for outright fabrication of facts, or it may be applied in situations with vague references or claims that do not appear relevant to larger environmental issues. In this case the sin of greenwashing precludes dialogue. There is no room for discussions of intentionality, context, or history.

Obviously many organizations have used green marketing in manipulative and unethical ways. Critics have listed a number of egregious examples in studies. But using the label to describe green advocacy as a genre of discourse is a convenient tool but one that shortchanges legitimate dialogue and casts doubt upon those organizations committed to positive environmental change. There should be an open quality to assessments of environmental advocacy that generate judgments that can be expressed in a detailed and analytical discussion.

The Power of Green Advocacy

"Weathercocks" offers many insights into the intellectual complexity and political roadblocks that define environmental politics. The report articulates useful suggestions for taking action, but a high degree of frustration accompanies the findings. One problem that inhibits the impact of environmental messages lies in how many observers view the environmental movement. In a world defined by mediated information, ideological fragmentation guides political discourse and in turn classifies the environmental movement as just one of many special interest movements in global politics. In this worldview, citizens see political urgency through the mediated lens of their own particular interest: poverty, gender rights, illegal immigration, abortion, education, animal rights, children, and so forth. But in pragmatic terms, the environment represents something larger, more consequential, and more fundamental than all other issues. The environment must become the *transcendent special interest* in political culture; without a healthy and functioning environment, there is little hope for accomplishing social and political justice in all other realms of life.

How can organizational rhetoric promote this transcendent view of the environment? Three perceptual changes will empower organizations to generate meaningful environmental messages in support of changing attitudes and behavior.

First, organizations should be viewed as legitimate forces of change in global society. Organizations are often more powerful than individual governments because of their ability to have a global relationship with multiple constituencies. In a postmodern culture, many activists see the organization as a force against change and as an enemy of the environment. But increasingly leaders in all types and sizes of organizations have asked for a place at the environmental policy table. For example, in *The Soul of a Business* Tom Chappell, founder of Tom's of Maine Toothpaste, described his commitment to making the world a better place through business practices. After attending divinity school, his entire orientation toward business changed. He learned that business lives in a larger world than profits and growth. Writing in 1994, Chappell predicted a paradigm shift in corporate leadership:

> Once confused about my priorities, I am now very clear. The ultimate goal of business is not profit. Profit is merely a means toward the ultimate aim of affirming the health and dignity of human beings and their families, affirming aspirations of the community, and affirming the health of the environment—the common good. If our air is polluted, our communities and people polluted, how can our businesses really prosper? (p. 202)

Chappell introduced morality into his theory of management, writing: "It is morally wrong for any boss to humiliate an employee, exploit his work, steal her ideas, pay him unfairly, or harass her sexually. It is not less morally wrong to violate the environment" (p. 205).

Giving organizations status, however, also means that they should be held accountable for their communication. Consumers, educators, activists, and others should evaluate green claims and share this information with their friends and colleagues. With the advent of cyber-information, the ability to access data and examine such claims is viable and provides a useful service for organizational stakeholders. For Chappell, the organizational commitment to goodness means "being held accountable to the values and expectations of the community—the common good" (p. 201).

Second, organizations should strive to integrate environmental issues in all aspects of presenting their mission. Stakeholders should demand that the organizations representing their interests address relevant and appropriate environmental challenges. Corporate reports are replete with environmental commitments by organizations, and green rankings provide stakeholders with valuable data that can be used in purchasing products, buying stocks, and seeking employment. For example, in 2009 *Newsweek* published a list of the 500 "Greenest Big Companies in America" with a detailed assessment of each company's environmental impact. The goal is "to open a conversation on measuring environmental performance—an essential first step toward improving it" (McGinn, p. 35). The magazine plans to conduct the survey every year using independent experts and seeking to improve its data base for greater accuracy.

Organizations will also achieve significant advantages by appealing to the social and environmental justice positions of potential employees. As the *Harvard Business Review* (2009) notes: "Recent research suggests that three-fourths of workforce entrants in the United States regard social responsibility and environmental commitment as important criteria in selecting employers." As a result, companies that "become sustainable may well find it easier to hire and retain talent" (Nidumolu, Prahalad, & Rangaswami, 2009, p. 9).

Third, organizations should assume rhetorical leadership in addressing the need for environmental change. They can achieve this position through advocacy connected to consumer education, creating partnerships, and adopting standards for message critique. A number of current examples illustrate how organizations are already assuming this rhetorical leadership.

In terms of education, The Cornucopia Institute publishes information related to organic farming, including an annual listing of organic milk produced in the United States and the dairies with the best and worst records (Cornucopia, 2010). *National Geographic* sponsors "The Green Guide," a website and blog with detailed information for consumers regarding the best and worst choices to make in terms of environmental impact as well as lots of other useful information to inspire sustainable living (*National Geographic*, 2010).

Partnerships can provide a powerful source of green advocacy. In 2008 an unusual partnership emerged when Clorox and the Sierra Club joined forces to market a line of environmentally friendly cleaning products. Recognizing that consumers wanted products that had little environmental impact, Clorox approached the Sierra Club for an endorsement. The club agreed to lend its logo to the product in exchange for a percentage of the sales of a product line called Green Works. After two years of sales, the Sierra Club has received $1.1 million for its sponsorship of ten different Green Works products. The products are not tested on animals, they use biodegradable ingredients and are packaged in recycled materials. Although a few members of the Sierra Club called this partnership an example of greenwashing and quit the group, organizational leaders said they examined the proposal and concluded it was in support of the organization's environmental agenda. The executive director of the Sierra Club, Carl Pope, observed that the organization joined with Clorox in order to help make "affordable and effective natural cleaning products available to millions of Americans. We are thrilled that since the launch of this partnership, the natural cleaning category has more than doubled—making a real impact" (Clorox, 2010).

To critique green advocacy, it is important to acknowledge that environmental messages presented by organizations are always about values, including issues related to nature, consumption and politics. A useful approach to assessing value claims comes from the work of Karl Wallace who authored an important study in 1955 titled "An Ethical Basis of Communication." Troubled by the blacklists of the 1950s, when people could be smeared by unsubstantiated claims, Wallace was also frustrated by his colleagues who believed the legitimate study of communication was only concerned with techniques of persuasion, not the ethical dimensions of persuasion. Wallace rejected this view, writing, "I believe that there are ethical standards which should control any situation in which speaker and writer endeavor to inform and to influence others" (1955, p. 2). In crafting his theory of communication ethics, Wallace looked toward the values Americans embraced in our founding documents. He created a list of "Four Habits" that could be used by both speakers and audiences to describe ethical communication practices. I believe Wallace's "Four Habits" provide a useful method to assess organizational advocacy and determine if a message is using green advocacy in an ethical manner. Specifically, Wallace wrote (1955):

- An ethical communicator must recognize his/her status as an expert on a given subject and act accordingly.
- An ethical communicator presents facts and opinions accurately and fairly.
- An ethical communicator must reveal sources of information for audiences.
- An ethical communicator must acknowledge and respect diversity of opinion.

By applying these standards to organizations using green marketing, we can make specific judgments about claims and if they should be given legitimacy or if they should be questioned as misleading or faulty. Individuals can apply these questions them-

selves or turn to third-party interest groups committed to evaluating the ethics of green branding. This might include volunteer
groups, activist groups, educational professionals, religious institutions, and non-profits as vehicles to seek information and make
assessments. An example of this kind of analysis of green marketing was published in 2004 by Ann Marie Todd, who analyzed the
green claims of health care products. She found that Burt's Bees,
Tom's of Maine, and The Body Shop avoided greenwashing "by offering detailed information on their websites, on their packaging,
and in their advertisements about their ingredients, philanthropic
commitments, and ethical practices" (2004, p. 99). In this way
Todd demonstrates that the green claims on these health care
products were ethically sound based on standards similar to those
posed by Wallace.

Faith Communities and Environmental Action

To achieve significant changes in consumption, Crompton notes
that while most theories of changing behavior have questionable
impact, if individual values can be linked to sustainable practices,
then change is more likely to emerge. In this regard he notes that
values are difficult to change and linked to individual identity. He
concludes that the environmental movement must find ways to
foster significant changes in individual behavior with issues related to values and to personal identity. An important sphere of connections between values and identity lies in faith communities. A
large and distinct group of Americans identifies strongly with a
faith tradition. A Pew Study notes that 78% of Americans consider
themselves Christians, and another 4% are affiliated with other
religions (Pew, 2010). Members of faith communities have often
been seen as the problem in fostering environmental awareness,
not as a source of support for sustainable practices. However,
there is strong evidence to suggest that American religious leaders
understand the need for significant changes in behavior and are
motivated to take that message to their members. Even conservative evangelicals have joined the crusade for climate change. Hundreds of these Christian leaders have joined together in The

Evangelical Climate Initiative and have crafted a statement favoring immediate and significant action to curb climate change. These leaders write:

> For most of us, until recently this has not been treated as a pressing issue or major priority. Indeed, many of us have required considerable convincing before becoming persuaded that climate change is a real problem and that it ought to matter to us as Christians. But now we have seen and heard enough to offer the following moral argument related to the matter of human-induced climate change (Evangelical, 2010).

This group is committed to using advocacy within the Christian community to promote significant changes that they divide into: Learn, Pray, Act.

The potential impact by turning to faith communities in the United States alone is staggering. The President's Advisory Council on Faith-Based and Neighborhood Partnerships (2010) has proposed a plan to engage the nation's religious organizations in addressing climate change and environmental concerns. The report notes, "The more than 370,000 houses of worship alone provide locations for information to be shared, training to take place, and modeling of best environmental practices to occur" (President's, p. 55). Christian leaders from across the spectrum from conservative to mainline to liberal have endorsed the need for change and are prepared to accept the need to motivate changes through education, partnerships, and critique. Writing in *Christian Century* about the impact of the Gulf oil slick, environmentalist Bill McKibben challenged his readers to use their churches to demand political change, "Churches are key to this work" he concluded (2010, p. 11). Religious studies professor Larry Rasmussen asked readers of *Sojourners* to live a "different faith and ethic" that remembers the true bottom line of faith: "The health and wellness of human beings is dependent on the wellness of our ecosystems" (2010, p. 26).

An example of how interdenominational faith communities can use green advocacy effectively can be found in the Iowa Interfaith Power & Light (Iowa Interfaith Power & Light, 2008). This group represents Christian, Jewish, Muslim and other faith communities

that seek to address climate change by providing education about
global warming, helping individuals and congregations reduce
their carbon footprints, advocating for sustainable energy, and
selling environmentally friendly energy products (Iowa Interfaith
Power & Light, 2008). The group sponsors events like "Cool Congregations" workshops, disseminates scientific information on climate change, and works as a clearing house for political activism.
Another group that seeks to unite multiple faith communities is
Faiths United for Sustainable Energy (FUSE). Noting that 80% of
the global population identifies with a major organized religion,
the group uses collective values to support environmental causes
such as climate change legislation, five million green jobs, and
elimination of coal as an energy source. FUSE uses education, advocacy, and partnerships to encourage changes in consumption
and to promote sustainable practices by individuals and groups,
including religious institutions and businesses (FUSE, n.d.).

For some social critics, these efforts by Christian groups to engage members through green advocacy may appear to be examples
of greenwashing, efforts to engage new members and show the relevance of the church in a postmodern world. At the core of these
efforts lies a moral argument that should engage all people, regardless of their religious beliefs. The moral argument speaks to
participation in environmental destruction and engages audiences
through the belief system that guides their own actions and values. As such, the faith community offers a rich source of citizen
energy that promotes working with other interested groups in a
partnership for achieving sustainable lives.

Conclusion

A culture of green advocacy can challenge our current degree of
collective complacency regarding environmental issues and inspire
significant changes in behavior. Organizations can emerge as fully
engaged partners in shaping the future and support Chappell's
belief that all organizations should manage for profit *and* for the
common good. Indeed, as Chappell writes, "Conserving the environment may be the most powerful force for change in America.

Wherever you live, people are trying to do something to help, sustain, and protect the environment" (p. 203). Green advocacy generates at least three benefits that can ignite large-scale changes in resource consumption and help people construct new identities based on a moral commitment to a larger cause than consumption and materialism.

First, green advocacy promotes social change through consumer action. Consumers become part of the cycle of changing the world's practices when they have legitimate and appropriate market choices. It creates a sense of partnership and provides a first step for developing greater sensitivity to environmental concerns. As Todd notes, green consumer goods are "ideological by their very nature and thus represent an ethics-based market with a consumer culture shaped by environmentally aware shoppers. Thus, eco-marketing constructs a complicated ethical identity for the green consumer" (2004, p. 88). Reporter Corby Kummer was shocked to hear that Walmart had created an organic food division called "Heritage Agriculture," which encourages farms within a day's drive of a Walmart warehouse to grow crops for local sales. Walmart claims it wants to revive local agricultural communities that were lost to centralized agribusiness in the past century. Kummer concludes, "It's not something you expect from Walmart, which is better known for destroying local economies than for rebuilding them" (2010, p. 40). Kummer purchased organic foods from a Boston area Whole Foods Grocery and from a local Walmart and invited friends to a tasting party. A number of his friends were surprised to learn they had endorsed Walmart produce in certain instances. Most importantly, Kummer concludes that while he is not convinced that Walmart is committed to organic agriculture philosophically, "I'm convinced that if it wants to, a ruthlessly well-run mechanism can bring fruits and vegetables back to land where they once flourished, and deliver them to the people who need them most" (2010, p. 41).

Second, green advocacy creates a model of innovation, sustainability and organizational values that can be shared and emulated by others. Organizations use advocacy to report success stories as well as set goals for the future. Addressing organizational stakeholders, such as consumers, employees, partners, and even competitors, some organizations have crafted annual sustainability reports. Johnson & Johnson health care products have a long history of being committed to social justice, and their 2007 sustainability report is filled with specific examples of their practices, goals, and contributions to protecting the environment. These include specific changes in environmental practices (reductions in carbon dioxide emissions and decreases in water consumption, for example) as well as future plans and a list of partnerships that includes working with the World Wildlife Fund, Harvard Medical School, the World Resources Institute, and the Environmental Protection Agency. These commitments helped Johnson & Johnson become number three in the *Newsweek* 500 greenest companies in the United States as well as win a number of other awards for its environmental agenda.

Another way organizational advocacy contributes to changing industry values is discussed in a 2009 study in *Harvard Business Review* which examined the sustainability initiatives of 30 large corporations. The authors offered the following conclusion:

> Our research shows that sustainability is a mother lode of organizational and technological innovations that yield both bottom-line and top-line returns. Becoming environmentally-friendly lowers costs because companies end up reducing the inputs they use....Indeed, the quest for sustainability is already starting to transform the competitive landscape, which will force companies to change the way they think about products, technologies, and processes, and business models. (Nidumolu, Prahalad, & Rangaswami, p. 2)

This study identifies multiple ways in which organizations have embraced sustainable practices, which may have appeared costly initially but in the long run have yielded both profit and positive steps toward sustainability.

Because communication is a consequential act and it continually shapes our social reality, we need to consider the positive act of embracing environmental values in a public forum. Richard Gregg (1971) wrote that rhetoric "has to do with constituting selfhood through expression." The verbalization of a public stance "takes on the aspects of both act and appeal, the two occurring simultaneously. Perhaps it is more accurate to say that rhetoric is part of act, and the adoption of a particular rhetorical stance is self-confirming and enhancing" (p. 75). Clearly this sense of selfhood has changed CEOs like Tom Chappell who use green advocacy to define their organization's practices and identity.

Third, green advocacy sets a standard of value to judge organizational claims and actions in the future. Organizations have sought to shape perceptions of stakeholders through many kinds of messages, often by addressing values. In offering an organizational commitment to a particular set of values, the organization is creating standards on which it will be judged by its stakeholders. In their study of the rise in values advocacy by organizations in the 1970s and 1980s, Denise Bostdorff and Steven Vibbert conclude:

> Values advocacy establishes a context of shared values that formulates bases of judgment which audiences used to appraise communicators and their future messages. As such, values advocacy not only is important in its own right, but also indispensable to other forms of persuasion. (1994, p. 154)

Organizations that turn to green advocacy should be evaluated on the standards implicit in their claims of being committed to environmental protection. As Bostdorff and Vibbert note, any person or group proclaiming a moral stance will be judged by others according to that same framework.

Organizations are essential for promoting good choices in a rhetorical democracy. As scholars of communication, we need to find ways to discuss organizational advocacy that empowers citizens, consumers, voters, and activists. Communication studies evolved in the twentieth century as a helping discipline, training people to be effective and ethical advocates as well as enlightened audience members. If we embrace cynicism about all those in positions of power, I am not sure we can find paths to make healthy decisions in the public sphere. Greenwashing is an attractive concept for people in a postmodern age. It affirms our nagging suspicion that organizations cannot be trusted and provides a convenient label for assessing green branding. But throughout American history, the hard work of making public policy for the good of the whole requires people who will work together in some capacity and strive for consensus. The global environmental crisis calls for continued efforts to find communication strategies that promote lasting and significant change. That's the beauty, and the challenge, of being engaged in our rhetorical democracy.

References

Bostdorff, D. and Vibbert, S. (1994). Values advocacy: Enhancing organizational images, deflecting public criticism, and grounding future arguments. *Public Relations Journal, 20*(2), 141–158.

Clorox Company (2010, March 1). Green works natural cleaners and Sierra Club celebrate two-year anniversary; doubling of natural cleaning category. Retrieved from http://investors.thecloroxcompany.com/releasedetail.cfm?release id=448538

Cornucopia Institute (2010). Dairy report and scorecard. Retrieved May 30, 2010 from: http://www.cornucopia.org/2008/01/dairy-report-and-scorecard/

Crompton, Tom (2008). "Weathercocks & Signposts: The Environmental Movement at a Crossroads." WWW-UK Strategies for Change. Retrieved from: www.valuesandframes.org/downloads

Evangelical Climate Initiative (2010). Retrieved May 30, 2010 from: http:// christiansandclimate.org/

FUSE. Faiths United for Sustainable Energy. (n.d.). Retrieved May 30, 2010 from: http://www.fusenow.org/

Gibson, D. (2009). Awash in green: A critical perspective on environmental advertising. *Tulane Environmental Law Journal, 22.*

Gregg. R. (1971). The ego-function of the rhetoric of protest. *Philosophy and Rhetoric 4* (1), 71–91.

Heath, R. (2006). Onward into more fog: Thoughts, on public relations' research directions. *Journal of Public Relations Research 18* (20), 93–114.

Iowa Interfaith Power & Light (2008). Retrieved from: http://www.iowaipl.org/ home.html.

Johnson & Johnson Sustainability Report (2007). Retrieved February 2, 2009 from: http://www.jnj.com/connect/caring/environment-protection/.

Kummer, C. (2010, March). The great grocery smackdown. *The Atlantic, 305*(2): 39–41.

Laufer, W. (2003). Social accountability and corporate greenwashing. *Journal of Business Ethics, 43*(3), 253–261.

McGinn, D. (2009, September 28). The greenest big companies in America. *Newsweek,* 35–54.

McKibben, B. (2010). What's worse than the Gulf oil leak? It's about the carbon. *Christian Century,* (June 1, 2010), pp. 10-11.

Mulch, B. (2009). The seventh sin. *Alternatives Journal, 35*(2), 40.

Munshi, D. and Kurian, P. (2005). Imperializing spin cycles: A postcolonial look at public relations, greenwashing, and the separation of publics. *Public Relations Review, 31,* 513–520.

National Geographic (2010). The green guide. Retrieved May 30, 2010 from http://www.thegreenguide.com/.

Nidumolu, R., Prahalad, C. K., and Rangaswami, M. R. (2009, September). Why sustainability is now the key driver of innovation. *Harvard Business Review*, 1–10.

Pew Forum on Religion & Public Life (2010, June 1). U.S. religious landscape survey. Retrieved May 30, 2010 from: http://religions.pewforum.org /affiliations

President's Advisory Council on Faith-Based and Neighborhood Partnerships (2010, March 11). A new era of partnerships, report of the recommendations to the president: 55–67. Retrieved May 30, 2010 from: http://www. whitehouse.gov/sites/default/files/microsites/ofbnp-council-final-report.pdf

Priesnitz, W. (2008). Greenwash: When the green is just veneer. *Natural Life*, 14–16.

Todd, A. M. (2004). The aesthetic turn in green marketing: Environmental consumer ethics of natural care products. *Ethics & the Environment, 9*(2), 86–102.

Voss, J. (2009). Actions speak louder than words: greenwashing in corporate America. *Notre Dame Journal of Law, Ethics, and Public Policy, 23*(2), 673–697).

Wallace K. R. (1955). An ethical basis of communication. *The Speech Teacher, 4* (1): 1–9.

CHAPTER FIVE

Environmental Risk Communication: Right to Know as a Core Value for Behavioral Change

Michael J. Palenchar

Bernardo H. Motta

In response to Tom Crompton's (2008) report, "Weathercocks and Signposts" this chapter discusses two areas: the role of right to know to achieve greater clarity on the values that motivate the environment movement and the role of right to know within narrative enactment while emphasizing intrinsic goals in environmental communications, including a brief discussion on intrinsic values and risk equity. This discussion is important to the study of environmental communication because the enforcement and success of the right-to-know policies depends almost solely in the ability of the public to receive, understand, process, use, and distribute information. Whether that information is wrapped up within a marketing approach—for good or bad—to creating behavioral change, the core value of right to know and a risk communication infrastructure that helps symbolically construct wise outcomes is fundamental to not only understanding but better addressing environmental challenges.

The Principle of Right to Know

The principles of the public's right to know, self-governance, and community involvement wrapped up in risk communication may constitute the core of environmental communication. In the last decades of the twentieth century, several authors, including Ulrich Beck and Anthony Giddens, started to see how the rapidly growing

complexity of modern social organizations made it effectively un-
feasible for any governmental institution to seriously address or
even begin to solve social problems, such as the collapse of our environmental system, relying solely on governmental apparatuses
(Beck, 1992; Giddens, 1991). At the same time, more and more authors in the social sciences and humanities have pointed out the
failure of exclusively market-based policies and communication in
providing just and desirable conditions to society as a whole. Tom
Crompton (2008) also questioned the sanity of a market-based approach and the sovereignty of consumer choice to environmental
communication, one built on more and different types of consumption as the solution to environmental challenges.

The context in which the twenty-first century society begun became known as what Beck called "the risk society." Beck (1992,
1999) proposed that, with the failure of social institutions to deal
with the broad concept of risk, society would need to turn more
and more to civic participation and self-governance in all stages of
government and society. For Beck, only through the inclusion of
the public in the decision-making process would governments be
able to prevent and ameliorate environmental problems. He also
defended the concept that well-informed local communities would
be more able to monitor and react to local risks. In Crompton's
(2008) report, he challenges the notion of information and information campaigns for effective behavioral campaigns related to
environmental challenges, building on the position "that we should
not expect information campaigns to create behavioral change" (p.
6). He added:

> Gone are the days when our focus lay on imparting information about
> environmental problems, in the forlorn hope that this alone would
> prompt mass pro-environmental behavioral change. (p. 8)

In response to Crompton's report, this chapter discusses the critical intersection between risk communication and the role played
by environmental risks in society, especially the concept of *right to
know*. The right-to-know approach to public policy—also known as
regulation through revelation—is based on the ideas of self-

governance and public participation in the decision-making process (Florini, 2007; Hamilton, 2005) and was made into a U.S. federal law in the Emergency Planning and Community Right-to-Know Act (EPCRA) of 1986. EPCRA has served as a model for more than 80 other countries since, for the better or the worse and was the first federal law in the United States to fully embrace the right-to-know approach to public policy.

This discussion is important to the study of environmental communication because the enforcement and success of the right-to-know policies depend almost solely on the ability of the public to receive, understand, process, use, and distribute information. Whether that information is wrapped up, sold as, and created within a marketing approach—for good or bad—to creating behavioral change, as is the critical focus of "Weathercocks and Signposts," the core value of right to know and a risk communication infrastructure that helps symbolically construct wise outcomes is fundamental to not only understanding but better addressing environmental challenges.

Specifically, this chapter addresses two areas of the report: (1) the role of right to know to achieve greater clarity on the values that motivate the environment movement; (2) the role of right to know within narrative enactment while emphasizing intrinsic goals in environmental communications, including a brief discussion on intrinsic values and risk equity, which is a fundamental concern to disadvantaged stakeholders inclusion in that decision-making process and being able to monitor and react to local risks. Access to information, especially useful information, and self-efficacy to address risk are challenging under ideal circumstances, an outright barrier when race and class power discrepancy, and other social and environmental justice issues, are involved. The challenge is in fully implementing right to know to address intrinsic values as the basis for environmental discussions, negotiations, policies, and outcomes.

Risk Society

Postindustrial societies can be characterized by their development from modern to risk societies (Beck & Holzer, 2007), where the locus of risk in these societies has shifted from nature-based, or outside risks, to industrial ones. These risks are rooted in the decision-making process of governmental and nongovernmental organizations and corporations. An organization's willingness to take calculable risks in order to advance, change, and create business is an inherent factor of today's Western societies (Beck & Holzer, 2007).

The principles of the public's right to know, self-governance, and community involvement, may constitute the core of EPCRA, but EPCRA's adoption was as firmly grounded in actual historic events as in ideological principles. In 1986 Beck published *Risikogesellschaft—Auf dem Weg in eine andere Moderne*, translated to English six years later as *Risk Society*. In it, Beck portrayed a constantly changing world in its process of modernization and free of the traditional gridlocks of the industrial society. Beck's thesis was that society was "witnessing not the end but the *beginning* of modernity—that is, of modernity *beyond* its classical industrial design" (p. 10). For Beck, that new stage of modernity is called *reflexive modernization*, one of his central theories: "The argument is that, while in classical industrial society the "logic" of wealth production dominates the "logic" of risk production, in the risk society this relationship is reversed" (p.12).

Although many other authors, like Rachel Carson (1962) for example, identified environmental problems as a serious threat to society, Beck's work created a new lens for the social sciences to study the concept of environmental risk as the central focus of analysis, replacing the role of social institutions. Beck (1992, 1999) also proposed that, with the failure of social institutions to deal with the broad concept of risk, society would need to turn more and more to civic participation and self-governance in all stages of government and society. For Beck, only through the inclusion of the public in the decision-making process would governments be able to prevent and ameliorate environmental problems.

At the heart of this misunderstanding or misperception be-tween acceptable and actual risk levels is a communication gap between those who decide which risks in a given context are al-lowable and those who bear the risk. Increasing technological complexity and fragmentation of organizational processes augment this gap between experts and risk bearers (Goldstein, 2005). This gap is further emphasized when race and class power discrepancy issues are involved.

Heath & O'Hair (2009) suggested that when a risk event be-comes manifest, it becomes a crisis. Researchers commonly ap-proach crises as objective events that can be understood and managed based on a set of best practices that are generalizable to many organizations. As alluded to by Renn (1992) in respect to risk conceptualizations, however, crises occur within the realm of social reality in which the meaning of events is socially construct-ed through language (Berger & Luckmann, 1966). Communication is the process of co-creating meaning. It follows that "crisis man-agement is a distinctly communicative phenomenon in which par-ticipants construct the meaning crises hold" (Hearit & Courtright, 2004, p. 205). Although the actual event is not a social construc-tion, the understanding, interpretation, and meaning of it are. Consequently, a crisis or risk is comprised of two dimensions: an actual and a perceived dimension. Environmental risks and crises are no different.

Viewing crisis in this form, according to Heath and Millar (2004), places risk communication in a rhetorical tradition, in which crises are contestable events. Those interpretations and narratives that best allow for co-creation of meaning or the shar-ing of a zone of meaning will frame the perceived dimensions of crisis (Heath & Millar, 2004). Crisis situations are furthermore a struggle for control that can be achieved on two levels: actions and communication. Not only the organization will offer interpreta-tions of the events but all stakeholders involved, and these differ-ent groups will present their stories in the marketplace of ideas, creating a rhetorical problem (Heath & Millar, 2004).

Nowhere is this presentation of risk narratives of environment among market place organizations more clearly pointed out and addressed than in "Weathercocks and Signposts." For-profit and environmental non-governmental organizations' (NGOs') marketing and communication techniques that address such massive-scale environmental challenges demonstrate how organizations build environmental narratives in the marketplace of ideas. Crompton's critical review of a marketing approach to creating behavioral change highlights the narratives that organizations will create and use, as opposed to dealing directly with two key points that are required to affect environmental change: right to know, and with that right to know "the scale of environmental challenges we confront demands a systematic engagement with this problem" (Crompton, 2008, p.6). This review covers such issues as the sovereignty of consumer choice, encouraging individuals to change their behavior for reasons of social status or financial self-interest, stressing the importance of small and painless steps, the consumption of more goods and services in response to environmental challenges—also known as green consumption—and the overall green consumption and consumerism.

Risk Communication

For this chapter, it is important to understand a critical intersection between risk communication and the role played by environmental risks in society of the late twentieth century and early twenty-first century. Cox (2006), for example, credited the importance of risk communication to the way in which it looks at the effectiveness of communication strategies for conveying information about health and environmental risks, the impact of cultural understanding of risk on the public's judgment of the acceptability of a risk, and the ways to develop more democratic methods to involve affected communities in evaluating risk.

According to the National Research Council (1989), risk communication is a means to open, responsible, informed, reasonable, scientific and value-laden discussion of risks associated with personal health and safety practices involved in living and working in close proximity to harmful activities and toxic substances—a concept that was early adopted by the Environmental Protection Agency.

According to Leiss (1996), risk communication started as a source-oriented approach based on the locale where risks are present. It then evolved to its present approach of communication based on shared social relations. "As such, there is no single psychology or sociology of risks....Risks are not necessarily selected and perceived due to their scientific merit or personal benefit, but out of a combination of social and cultural factors, denotative and connotative reasons" (Palenchar, 2008, p. 3). Lundgren and McMakin (2004), for example, listed 12 approaches to the study of risk communication: communication process, the NRC's approach, mental models, crisis communication, convergence communication, three-challenge, social constructionist, hazard plus outrage, mental noise, social network contagion, social amplification of risk, and social trust.

Cox (2006) argued that "the field of environmental communication arises at a moment of conjunctural crisis, defined in not insignificant ways by human-caused threats to both biological systems and human communities, and also by the continuing failure of societal institutions to sufficiently engage these pressures" (p. 7). Concerning communication related to complex environmental issues, for example, it is understandable that risk messages can be confusing for risk bearers; they come from a variety of sources that involve multiple parties and often reflect competing scientific conclusions. Experts and regulatory agencies often operate on the assumption that they and their audiences share a common framework for evaluating and interpreting risk information (Heath, Palenchar, Protheau, & Hocke, 2007), even if that common framework is one that reflects intrinsic values as advocated by Crompton.

Symbolic anthropologist Mary Douglas (1992) and her colleagues such as Steve Rayner (1992) have argued that society organizes on the single premise that the rationale for forms of human associations is the collective management of risk; risks and crises are political events, as well as scientific ones, where facts blend with values and policy preferences. It is here where Crompton's argument for communication and action based on intrinsic values has a role, which will be discussed later in this chapter. Crompton and others who are involved in trying to change fundamental and systematic environmental practices that are devastating to our world are concerned about the quality of communication that leads to, results in, and emanates from scientific investigation and debate to ultimately affect systemic behavioral change.

Palenchar and Heath (2006) offered several ethical commitments to strategic risk communication based on the symbolic and cultural perspectives of risk that address this clash of perspectives, including: ideas and meaning count, the appropriate utilization of narrative enactment, acknowledging uncertainty, building trust through community outreach and collaborative decision-making, responsible advocacy, transparency, and the good organization must communicate well. These normative assumptions were advocated as important requirements for organizations that want to address and respond to the calls concerning risk related to systemic environmental challenges—key for risk communication if it is to add value to society.

Environmental Risk Movement

Despite the fact that by 1958 every American state had experienced cases of toxic contamination of ground water (Fletcher, 2003), it was only in 1962, with Rachel Carson's book *Silent Spring*, that environment and public health were stirred together once again. Carson's book resuscitated the discussion about environmental degradation by showing how human actions can have a "domino effect" on nature. *Silent Spring* was the seed for the grassroots effort that later included the Anti-toxic Movement, one of the most effective aspects of the new and reconfigured environment movement (Brulle, 1996; Mowrey & Redmond, 1993).

The Anti-toxic Movement helped increase the public outcry for regulations on the use of toxic and hazardous materials by industry and became instrumental (although not alone) in the creation of a series of laws and regulations—among them the Resource Conservation and Recovery Act, the Toxic Substance Control Act, the Clean Air Act, and the National Environmental Policy Act— and even the birth of the Environmental Protection Agency (EPA) in 1970 (Cole & Foster, 2001). In the end of the 1970s and early 1980s, a series of events put together environmental concerns, civil rights, and social justice concepts that became known simply as environmental justice (Taylor, 1997).

Initially, what is known now as the Environmental Justice Movement (EJM) was actually a myriad of different smaller movements, like the Anti-toxic Movement (now called the Environmental Health Movement), a few Not-In-My-Back-Yard (NIMBY) actions, the People of Color Environmental Movement and a mix of influences from the Civil Rights Movement, Native American struggles, the Labor Movement, traditional environmentalists, and academics (Cole & Foster, 2001). The EJM began to emerge from geographic areas that were primarily populated by people of color or lower economic status that had been affected by air, water, and soil pollution (Cable & Cable, 1995; Gotleib, 1993; Pellow & Brulle, 2005). These sometimes organized groups aimed to force corporations and governments to clean up and compensate local communities for the spill of toxic chemicals in the soil, water, and air and to impede the allocation of high-risk industrial plants in ethnic or poor neighborhoods among other issues.

The EJM differs from the larger environmental movement that had been active in the United States since the late 1800s (Taylor, 1997). Changing the locus of the environmental struggle from the wilderness to populated areas, the EJM worked from the grassroots level to include where people live, work, and play as well as how these things interact with the physical and natural worlds (Bullard, 2005). The EJM was born as a "political response to the deterioration of the conditions of everyday life as society reinforces social inequalities" and has sought to "redefine environmentalism as much more integrated with social needs of human populations" (Pellow & Brulle, 2005, p. 3).

The concept of environmental justice is an important part of the environmental risk communication model. Among the most important issues in contemporary environmental and risk communication, environmental justice addresses perceived race and class inequities in the distribution of environmental risks. Community residents who live near or work at manufacturing facilities that produce potentially hazardous and toxic materials are sensitive to the fairness and equity of risk distribution and to the resulting environmental and aesthetic implications of activities of such facilities (Bullard, 1994; Webler & Tuler, 2006).

The equity consequences of risk regulation have become a formal component of government evaluations of risk and environmental policies (Viscusi, 2000). For example, Executive Order 12,866, the chief legal instrument governing agency policy analysis, states that agency regulations should maximize net benefits and then address distributive impacts and equity issues. For example, Executive Order 12,898, the Environmental Justice Order, states that to "the greatest extent practicable and permitted by law...each Federal agency shall make achieving environmental justice part of its mission by identifying and addressing, as appropriate, disproportionately high and adverse human health and environmental effects of its programs, policies and activities on minority populations and low-income populations in the United States."

Risk equity is a fundamental concern to disadvantaged stakeholders' inclusion in that decision-making process and being able to monitor and react to local risks. Risk equity concerns are at the cornerstone of right-to-know policies. Access to information, especially useful information, and self-efficacy to address risk are challenging under ideal circumstances; an outright barrier when race and class power discrepancy issues are involved. Theories or practical applications of environmental communication that fail to acknowledge or exaggerate the power that lower socioeconomic communities have to protect themselves may work more in favor of the status quo. Ultimately right to know is about advocating social justice communication for issues, risks, and crises and moving beyond the sovereignty of consumer choice as one of the guiding communication mechanisms of environmental communication.

The perception of risk or fair allocation of risk is among the numerous motivators people use when deciding whether a problem exists and deserves their attention. Once in the face of a risk situation, people may opt to make personal responses or to collaboratively seek collective solutions by engaging in public policy struggles (Skanavis, Koumouris, & Petreniti, 2005). This carries two core concerns of both environmental and risk communication: the existence of real risk or actual problem and the public perception about them, which may empower or discourage communities to act. Hadden (1989) wrote that the role of risk communication, very much like environmental communication, is to function as a learning system that provides citizens with an understanding of risks, hazards, and health issues, but that it also works to help people use this knowledge in the political arena, improving public regulation and corporate practices. "If citizens are to take action based on information, they must understand it," (p. 137).

Right to Know

The right to know, as an idea, evolved in two distinct periods. In its pre-conceptual form, it was a right to be educated about the purposes of government and to be informed about the actions of government as formulated by John Milton, William Bollan, Thomas Jefferson, Thomas Paine, George Mason, John Wilson, and James Madison, among others. In its contemporary understanding, especially following Hadden's (1989) definition, the people's right to know is a mechanism to empower people through education and knowledge so they can watch over their governments and industries and improve their lives.

Since the earliest debates on American independence, the ideas of freedom and self-governance were in the forefront of the revolutionary discussions (Rabban, 1985). For Jefferson, the concept of a right to know was more than a simple right to access information; it included the right to be educated about the functions of the government, so people (through the press) could watch and control the actions of government. In a letter to Edward Carrington in January 1787, Jefferson wrote:

> I am persuaded myself that the good sense of the people will always be found to be the best army. They may be led astray for a moment, but will soon correct themselves. The people are the only censors of their governors: and even their errors will tend to keep these to the true principles of their institution. To punish these errors too severely would be to suppress the only safeguard of the public liberty. The way to prevent these irregular interpositions of the people is to give them full information of their affairs thro' the channel of the public papers, & to contrive that those papers should penetrate the whole mass of the people. The basis of our governments being the opinion of the people, the very first object should be to keep that right; and were it left to me to decide whether we should have a government without newspapers or newspapers without a government, I should not hesitate a moment to prefer the latter. But I should mean that every man should receive those papers, and be capable of reading them. (Epps, 2008, p. 74)

After the failure to expressly include a right to know in the U.S. Constitution, the first attempts to establish a right to know as law came from a few legal scholars and journalists during the 1950s. Harold Cross, a retired attorney for a newspaper, made the term "right to know" popular with the publication of *The Public's Right to Know*, in which he presented his findings of a study about access to government information, focusing especially on the Administrative Procedure Act (APA) of 1946 for the American Society of News Editors (Cross, 1953). Section 3 of APA made some matters of official record available to "persons properly and directly concerned" (APA, 1946). However, Cross discovered that the vague text of the law and the poorly defined exemptions were used by agencies to deny access to information and not to make it available. Cross's findings prompted a series of analyses and essays on the right to know, especially during the 11 years of hearings for the Freedom of Information Act of 1966 (FOIA), an amendment to APA signed by President Lyndon Johnson.

Three years after of the enactment of FOIA, another statute carrying a different version of right to know was enacted as a direct consequence of the efforts of the Environmental Movement:

the National Environmental Policy Act (NEPA) of 1969. However, the right to know in NEPA, like in FOIA, was still not fully developed.

> [W]hile NEPA was certainly a groundbreaking statute in the broad public "right-to-participate" sense, it does not clearly fit into the narrower "right-to-know" category which is more often thought of as a scheme in which polluters are required to disclose their processes and other information. The first true right-to-know provision in a substantive environmental statute would not become law until 1986. (Jacobson, 2003, p. 348)

The ideological concept of the right to know, although defended by different people at different times, has never been fully recognized by the Supreme Court or clearly stated in a federal law until the Emergency Planning and Community Right-to-Know Act (EPCRA) of 1986. It's important to highlight both the *preventive* and *participatory* characteristics of a right to know in EPCRA as it represents a good step to empower citizens to act directly in the policymaking process and take part in the decision-making. Thus they differentiate EPCRA from previous laws, like the Freedom of Information Act, that brought a right to petition the government for information but not a right to know with the preventive and participatory characteristics. Furthermore, EPCRA also included private corporations into the mix of entities that needed to make the information available, recognizing them as part of the group of institutions that directly affected the public interest and, therefore, needed to provide checks and balances on the activities that put public health at risk.

Many others have taken it a step further by suggesting an actual "means" to include citizens in the decision-making process beyond the educative component. According to Hadden (1989), risk communication is a democratic tool for participatory citizenship where the right to know is also the right to take part in the process. Thus, the role played by environmental risk communication is fundamental to understanding the development of the right to know in EPCRA as a different approach to policy that goes beyond the previous command-and-control and market-based approaches.

The structure of EPCRA differs from the structure of most laws as it follows a tripartite-dialogical approach between public, industry, and government based on information, as Bolstridge (1992) wrote:

> Most importantly, EPCRA did not require the EPA to evaluate or interpret the information collected, but only to make the information available in its original form and through some specific types of reports and analyses. The law is based on the premise that it is the responsibility of the public to ask questions and of facilities to address questions as they arise. (p. 3)

The contribution of EPCRA to the policy process after the failure of previous legislations was to provide a way for society to achieve a necessary social goal—the reduction of environmental and health risks related to toxic and hazardous materials—despite the inefficiency of government. According to the main author of EPCRA, James Florio (1985), EPCRA, as an amendment to the Comprehensive Environmental Response, Compensation, and Liability Act (CERCLA), was a provision that should be easy for industry to comply with, inexpensive for government to administer, and empowering to citizens. The idea was that, once the information was compiled and published, scientists and citizens' groups could use it to make connections and see which facilities were in compliance with the earlier federal environmental legislations and the local and state laws and regulations.

Hadden (1989) developed a contemporary four-level concept of right to know applied to environmental risks involving toxic and hazardous chemicals that provides a solid summary of the concept. The *basic level* has the purpose of ensuring that citizens can find information about chemicals and holds the government accountable for ensuring that data are created and available. The *risk reduction level* aims to reduce risks from chemicals, "preferably through voluntary industry action but also by government if necessary" (p. 17). The *better decision-making level* allows citizens to participate in the decision-making process about the appropriate levels of hazardous materials in their communities. Finally, the *alter balance of power level* empowers citizens to participate in the decision-making process at the same or higher level than government and industry. Ultimately risk communication is educational and culture-forming processes much in line with the understanding of the role of informed and educated citizenry.

Community right to know might be the cornerstone for expanding knowledge of theoretical and practical important aspects of environmental communication. Community right to know, as both an ethical communication and business model, addresses other chapters in the book related to green campaigns and the incorporation of green values in an organization, provides a sounding board for the critique of marketing approaches of environmental issues, and fits in the overall construction of marketplace advocacy. Community right to know is implicit in which organizational voices are being heard on environmental issues, what those voices are saying, and to what effect, as Crompton calls for in "Weathercocks and Signposts." Right to know addresses core values and identities as opposed to marketing approaches to organizational environmental messaging that Crompton criticizes. Ultimately right to know is about advocating social justice communication for environmental issues, risks and crises, and moving beyond the sovereignty of consumer choice as one of the guiding communication mechanisms for the environmental movement.

Discussion

With a baseline knowledge of risk society, risk communication, the environmental risk movement, all wrapped up in the core value of right to know, the following discussion addresses several issues raised by Crompton in "Weathercocks and Signposts" in an effort to push the communication field into being a partner in addressing communication challenges related to positively and significantly addressing environmental challenges.

Right to know as a core value of environmental communication addresses all eight practical steps raised by Crompton (2008, pp. 35–36) in "Weathercocks and Signposts," which include: (1) achieve greater clarity on the values that motivate the environment movement; (2) emphasize intrinsic goals in environmental communications; (3) begin to deploy a broader vocabulary of values in policy debates; (4) find common ground with development agencies on these values; (5) help responsible businesses think beyond "the business case for sustainable development," (6) highlight the way in which the marketing industry works to manipulate our motivations; (7) work to support and embolden public figures in the course of articulating intrinsic values in public discourse; and (8) identify and promote mechanisms to make public affinity for nature more salient.

However, this discussion will specifically address two areas: (1) the role of right to know to achieve greater clarity on the values that motivate the environment movement; and (2) the role of right to know within narrative enactment while emphasizing intrinsic goals in environmental communications, including a brief discussion on intrinsic values and risk equity. The latter is a fundamental concern to disadvantaged stakeholders looking for inclusion in that decision-making process and being able to monitor and react to local risks.

Achieve Greater Clarity on the Values That Motivate the Environment Movement

The core value of right to know is about clarity of not only issues but also values that underlie and motivate behavior. Crompton (2008) argued that there is "little consistency in the values that the environmental movement reflects" (p. 35) with different values being highlighted with different audiences. Consistency and clarity are admirable goals of risk and environmental communication but hardly attainable with such diverse objectives and perceptive risk challenge.

The focal issue is not the objective science and related conclusive research findings about potential environmental catastrophes such as the collapse of fish stocks, deforestation, and pollution of watercourses. The focal issue related to communication is in risk controversies and is therefore the social and/or institutional uncertainty surrounding the decision, which is then itself amplified by the existence of scientific or technical uncertainty. As Heath et al.

(2007) noted in their argument about risk and crisis communication and the environment:

> [U]nder this rationale, stakeholders to the environments can focus their advocacy more efficiently to foster identification and avoid misidentification. In this context, environmental communication ought to help conservation stakeholders, allied and foes, to define the real issues to be discussed in public forum, hence inscribing its advocacy in the community's interest. (p. 43)

Because risks are defined in probabilistic terms, uncertainty can be viewed as "the lack of attributional confidence about cause-effect patterns" (Albrecht, 1988, p. 387). In the face of problems of attribution, humans spontaneously bridge ambiguities, characteristic of uncertain terrains, by temporizing their classifications of the world (Burke, 1945, 1961). In this way, the uncertainties straining the essence of their commitments become diluted in their narratives, which in turn become the primary perspectives from which they evaluate the world (Heath et al., 2007).

Risk-managing organizations, such as governments and NGOs, set different approaches when facing the same risk situations. According to Holling (1986), these strategies relate to various interpretations of ecosystem stability or the "different concepts people have of the way natural systems behave, are regulated, and should be managed" (p. 296). The particulars of these reflections correspond to the broader statement that in some cultures the orientation of man to nature falls into positions ranging from man's control over nature to man's subjugation by nature to man's harmony with nature (Kluckhohn & Strodtbeck, 1961).

However, when these choices involve the extremes of uncertainty, the interest of the community ought to override them. This way, cultural risk theory is capable of "illuminating and appreciating the complex connection between formative pressures exerted by social environment and the culture-creative responses of individuals to those pressures" (Gross & Rayner, 1985, p. xiv) towards the goal of "identify[ing] both the cultural variations that can block coordination and the cultural constructions that can increase perceptions of uncertainty" (Gerlach & Rayner, 1988, p. 16). Greater clarity is necessary but acknowledging ambiguity and uncertainty in science, let alone the political process of environmental decision-making, is a key component of this clarity.

Emphasize Intrinsic Goals in Environmental Communications

Intrinsic values are discussed throughout "Weathercocks and Signposts" as oriented toward personal growth, relationships, and community involvement. This narrative is one of many competing voices in the discourse of environment, along with the notion of equating pro-environmental behavior with self-sacrifice, sovereignty of consumer choice, encouraging individuals to change their behavior for reasons of social status or financial self-interest, stressing the importance of small and painless steps, and green consumption.

Crompton (2008) highlighted research that shows that individuals reporting higher subjective well-being also exhibit more pro-environmental behavior and that this is mediated by intrinsic values. According to Crompton, "But this approach does nothing to begin to dismantle the more systematic problems arising from the misplaced perception that happiness is best pursued through the acquisition of material objects" (p. 30). He argued that it is better to frame these communications in terms of a set of intrinsic goals. "To do so will simultaneously serve to increase the legitimacy of public debate framed in terms of such goals, and may well lead to more energetic and persistent audience engagement with environmental issues" (p. 35). (See Section 3.5 of "Weathercocks and Signposts" for more details on intrinsic goals in environmental communication).

From this perspective, and from the perspective of right to know, a marketing approach, one of the many narratives of environmental communication, may be playing too large a role in the discourse of addressing environmental challenges. Risk cultures offer different narratives that frame their perspective on a specific risk. These different narratives are presented in the marketplace of ideas (Heath, 1992, p. 18) and are contested through rhetoric.

"With rhetoric, people make collective decisions and form policy for the public good" (Heath, 2001, p. 37).

Narratives that are accepted through dialogue are integrated into the identity of a society (Fisher, 1987). Narratives are used to create, maintain, and continue the interpretation needed for stabilizing the distribution of power within a society. Power is exercised by those groups best able to frame their interests as those of other groups (Heath, 1994, p. 66). In any time of risk, the interpretations/narratives offered to frame and explain uncertainty favor those of the empowered groups (Heath et al., 2007).

In democracies, different groups share power. Narratives are used by these groups as an ideological force articulating a structure of significance that supports their privileged status. Thus the varying narratives used by an eclectic range of organizations are the natural result of dialogue and discourse on such a complex yet economically and culturally important science and life topic. For-profit organizations' willingness to engage in green marketing campaigns based on the sovereignty of consumer choice, as thoroughly described by Crompton, is a meta-narrative designed to preserve current lifestyles based on consumerism, even if this consumerism shifts toward a greener lifestyle, regardless of the implications for our environment—good or bad or indifferent.

According to Crompton's (2008) research, consumers' self-identity in relationship to environmental challenges is being motivated by the marketing industry. "[O]ur material possessions come to define who we see ourselves as being, and who we want to be seen to be" (p. 7). The challenge for Crompton and others is developing a narrative based on those intrinsic values that ring true with various stakeholders' constructed identities and their constructed sense of risk related to the environmental risks.

At the same time the narrative of regulatory agencies and other governmental agencies is to minimize their role in the debate with the hope of regulation that is difficult to pass while emphasizing the role of the private sector and even consumers in their purchasing and service patterns. As Crompton (2008) so passionately noted, "Focusing on private-sphere behavior change may serve as a dangerous distraction from the serious business of getting in place policy frameworks that are sufficiently ambitious to address these environmental challenges systematically" (p. 13).

A very important section of Crompton's report documents an emerging consensus on pro-environmental change strategies or a marketing approach. Crompton draws on a number of recent reports published by UK-based think tanks, communication consultancies, and organizations. Two of the six ways this marketing approach is characterized have direct implications for this discussion. This includes a reliance being placed on 'small steps', often in the expectation that these will lead individuals to engage in more significant behavioral changes; and a particular emphasis is being placed on marketing green products or services—also called green consumption.

A reliance on small, attainable actions has been at the core of research and application of right to know and risk and crisis communication. Right to know argues that not only is risk information critical to discourse as well as access and power to affect change but that attainable actions are the starting point to affecting that change; self-efficacy has been a fundamental variable in risk communication research.

Messages that contain elements of self-efficacy can help restore a sense of control in an uncertain situation (Seeger, 2006). Within social media and risk situations, for example, user empowerment, such as that provided through active participation in wikis and other social networking sites 'can generate a sense of ownership' (Colley & Collier, 2009, p. 35), and risk observers can 'leave [the sites] recognizing themselves as members of a public' (Center for Social Media, 2009). In another example, Sutton, Palen, & Shklovski (2008) found that during crises individuals felt 'a need to contribute, and by so doing, were better able to cope with the enormity of the situation' (p. 5).

Crompton (2008) is a bit surprised that small-steps by individuals and organizations might become the norm. "[I]f the belief that we ought to be able to address problems such as climate change—without this entailing fundamental life-style changes—actually sticks, this may serve to build resistance to the far-reaching government interventions that are actually needed" (p. 13).

Crompton's (2008) argument is that "such approaches may actually serve to defer, or even undermine, prospects for the more far-reaching and systematic behavioral changes that are needed" (p. 5) while highlighting the frequent ineffectiveness of narrative, dialogue, and overall rhetoric. However, rhetoric focuses on strategic communication influences that at their best result from or foster collaborative decisions and co-created meaning that align stakeholder interests. In theory and practice external organizational rhetoric weighs self-interest against others' enlightened interests and choices, and organizations as modern rhetors engage in discourse that is context relevant and judged by the quality of engagement and the ends achieved thereby. In theory and practice external organizational rhetoric weighs the relationship between language that is never neutral and the power advanced for narrow or shared interests (Palenchar, 2011).

Conclusion

So we go back to the beginning and reflect on something Crompton (2008) wrote in his executive summary:

> This report constructs a case for a radically different approach. It presents evidence that any adequate strategy for tackling environmental challenges will demand engagement with the values that underlie the decisions we make—and, indeed with our sense of who we are. (p. 5)

Our values, and our sense of who we are as individuals, organizations, and as a global society confronted with environmental challenges that are at times incomprehensible and on such a large scale that most individuals can't imagine how to grasp solutions, are socially and symbolically constructed from a sense of those values of all organizations engaged in discourse around this topic—including those who view a marketing-based approach as the best solution for a market-based economy and society.

Right to know as a core value of environmental communication and policy also brings us back to the beginning—an open exchange

of ideas and information, both scientifically objective and culturally interpretivistic, that acknowledges and appreciates the socially
constructed nature of knowledge and related actions to address
environmental challenges. While Crompton (2008) accurately argues that "it seems clear that approaches to addressing the problems of consumption must first engage in the underlying motives
that drive consumerism" (p. 20), the only long-term solution to our
environmental challenges is for consumers and policy makers to
understand these motives and the only way for that to start hap-
pening is dialogue based on right to know—both of information
and the core values that affect how we use that knowledge and
how we apply our resources and social and political capital.

As Crompton (2008) so aptly summarized, "Any adequate response to these challenges will require a re-examination of the relationship between people, and between people and the natural
environment. The sooner we embark on this re-examination, the
quicker we can move to institute fundamental changes that today's
environmental crisis demands" (p. 8). Environmental communication, like risk communication, integrates a focus on the probative
force of various facts in the context of symbolic processes that result from the dynamics of a functional or dysfunctional communication infrastructure (Heath et al., 2007). The role of risk
communication concerning such significant environmental challenges as discussed in "Weathercocks and Signposts" should be to
increase the quality of enlightened decision-making so that societies can be more fully functional in their identification, assessment, and management of risks. According to Heath et al., this
requires the shaping and application of a functional set of shared
principles that have scientific validation and reflect the cultural
dimensions of a risk society. "It is becoming increasingly clear that
the main product of environmental and risk communication is not
informed understanding as such, but the quality of the social relationship it supports, becoming a tool for communicating values and
identities as much as being about the awareness, attitudes, and
behaviors related to the risk itself" (p. 46).

References

Administrative Procedure Act (1946), 3, 5 U.S.C. 1002.

Albrecht, T. L. (1988). Communication and personal control in empowering organizations. In J. A. Anderson (Ed.), *Communication yearbook* (Vol. 11, pp. 380–390). Newbury Park, CA: Sage.

Beck, U. (1992). *Risk society: Towards a new modernity* (M. Ritter, Trans.). London: Sage.

Beck, U. (1999). *World risk society.* Cambridge, UK: Polity Press.

Beck, U., & Holzer, B. (2007). Organizations in world risk society. In C. M. Pearson, C. Roux-Dufort, & J. A. Clair (Eds.), *International handbook of organizational crisis management* (pp. 3–24). Los Angeles, CA: Sage.

Berger, P.L., & Luckmann, T. (1966). *The social construction of reality. A treatise in the sociology of knowledge.* Garden City, NY: Anchor Books.

Bolstridge, J.C. (1992). *EPCRA data on chemical releases, inventories, and emergency planning: A guide to the information on industrial facilities and chemicals available under the Emergency Planning and Community Right-To-Know Act.* New York: Van Nostrand Reinhold.

Brulle, R.J. (1996). Environmental discourse and social movement organizations: A historical and rhetorical perspective on the development of U.S. environmental organizations. *Sociological Inquiry, 66*(1), 59–83.

Bullard, R.D. (2005). Environmental justice in the twenty-first century. In R. D. Bullard (Ed.), *The quest for environmental justice: Human rights and the politics of pollution* (pp. 19–42). San Francisco: Sierra Club.

Bullard, R. D. (1994). Overcoming racism in environmental decision making. *Environment, 36*(4), 10–20.

Burke, K. (1945). *A grammar of motives.* Berkeley: University of California Press.

Burke, K. (1961). *A rhetoric of religion.* Berkeley: University of California Press.

Cable, S., & Cable, C. (1995). *Environmental problems grassroots solutions: The politics of grassroots environmental conflict.* New York: St. Martin's Press.

Carson, R. (1962). *Silent spring.* Boston: Mariner Books.

Center for Social Media. (2009). *Public media 2.0: Dynamic, engaged publics.* Retrieved from http://www.centerforsocialmedia.org/future-public-media/ documents/white-papers/public-media-20-dynamic-engaged-publics

Colley, K. L., & Collier, A. (2009). An overlooked social media tool? Making a case for wikis. *Public Relations Strategist, 34–35.*

Clean Air Act of 1970, 42 U.S.C. §7401 et seq.

Cole, L. W., & Foster, S. R. (2001). *From the ground up: Environmental racism and the rise of the environmental movement.* New York: New York University Press.

Comprehensive Environmental Response, Compensation, and Liability Act of 1980, 42 U.S.C. §9601 et seq.

Cox, R. (2006), *Environmental communication and the public sphere*. Thousand Oaks, CA: Sage.

Crompton, T. (2008, April). "Weathercocks and signposts: The environmental movement at a crossroads." WWF-UK. Retrieved from www.valuesandframes.org/downloads

Cross, H. (1953). *The public's right to know*. New York: Basic Books.

Douglas, M. (1992). *Risk and blame: Essays in cultural theory*. New York: Routledge.

Emergency Planning and Community Right-to-know Act of 1986. 42 U.S.C. 11001 et seq., partially codified at 42 U.S.C. §§ 9601-9675 (1988).

Epps, G. (2008). *The First Amendment: Freedom of the press: Its constitutional history and the contemporary debate*. Amherst, NY: Prometheus Books.

Fisher, W. R. (1987). *Human communication as narration: Toward a philosophy of reason, value, and action*. Columbia: University of South Carolina Press.

Fletcher, T. H. (2003). *From Love Canal to environmental justice: The politics of hazardous waste on the Canada-U.S. border*. Toronto: Broadview Press.

Florini, A. (Ed.). (2007). *The right to know: Transparency for an open world*. New York: Columbia University Press.

Florio, J. J. (1985, June 2). Chemical Russian roulette. *The New York Times*, p. 26.

The Freedom of Information Act § 5 U.S.C. § 552 (1966).

Gerlach, L.P., & Rayner, S. (1988). Culture and the common management of global risks. *Practicing Anthropology, 10*(3/4), 15–18.

Giddens, A. (1991). *Modernity and self-identity*. Palo Alto, CA: Stanford University Press.

Goldstein, B D. (2005). Advances in risk assessment and communication. *Annual Review of Public Health, 26*, 141–163.

Gotleib, R. (1993). *Forcing the spring: The transformation of the environmental justice movement*. Washington, DC: Island Press.

Gross, J., & Rayner, S. (1985). *Measuring culture*. New York: Columbia University Press.

Hadden, S.G. (1989). *A citizen's right to know: Risk communication and public policy*. Boulder, CO: Westview Press.

Hamilton, J.T. (2005). *Regulation through revelation: The origins, politics, and impacts of the toxic release inventory program*. Cambridge, MA: Cambridge University Press.

Hearit, K.M., & Courtright, J.L. (2004). A symbolic approach to crisis management: Sears defense of its auto repair policies. In D. P. Millar & R. L. Heath (Eds.), *Responding to crisis. A rhetorical approach to crisis communication* (pp. 201–212). Mahwah, NJ: Lawrence Erlbaum.

Heath, R.L. (1992). The wrangle in the marketplace: A rhetorical perspective of public relations. In E.L. Toth & R.L. Heath (Eds.), *Rhetorical and critical approaches to public relations* (pp. 17–36). Hillsdale, NJ: Lawrence Erlbaum.

Heath, R.L. (1994). *Management of corporate communication. From interpersonal contacts to external affairs.* Hillsdale, NJ: Lawrence Erlbaum.

Heath, R.L. (2001). A rhetorical enactment rationale for public relations: The good organization communicating well. In R.L. Heath (Ed.), *Handbook of public relations* (pp. 31–50). Thousand Oaks, CA: Sage.

Heath, R.L., & Millar, D.P. (2004). A rhetorical approach to crisis communication: Management, communication processes, and strategic responses. In D.P. Millar & R.L. Heath (Eds.), *Responding to crisis: A rhetorical approach to crisis communication* (pp. 1–18). Mahwah, NJ: Lawrence Erlbaum.

Heath, R.L., & O'Hair, H.D. (2009). The significance of crisis and risk communication. In. R.L. Heath & H.D. O'Hair (Eds.), *Handbook of risk and crisis communication* (pp. 5–30). New York: Routledge.

Heath, R.L., Palenchar, M.J., Proutheau, S., & Hocke, T. (2007). Nature, crisis, risk, science, and society: What is our ethical responsibility? *Environmental Communication: A Journal of Nature and Culture, 1*(1), 34–48.

Holling, C.S. (1986). The resilience of terrestrial ecosystems: Local surprise and global change. In W.C. Clark & R.E. Munn (Eds.), *Sustainable development of the biosphere* (pp. 292–320). New York: Cambridge University Press.

Jacobson, J.D. (2003). Safeguarding national security through public release of environmental information: Moving the debate to the next level. *Environmental Law, 1*(9), 327–397.

Kluckhohn, F.R., & Strodtbeck, F.L. (1961). *Variations in value orientations.* Evanston, IL: Row, Peterson.

Leiss, W. (1996). Three phases in the evolution of risk communication practice. *Annals of the American Academy of Political and Social Science, 545,* 85–94.

Lundgren, R.E, & McMakin, A.H. (2004). *Risk communication: A handbook for communicating environmental, safety, and health risks* (3rd ed). Columbus, Ohio: Battelle Press.

Mowrey, M., & Redmond, T. (1993). *Not in our backyard: The people and events that shaped America's modern environmental movement.* New York: William Morrow.

National Environmental Policy Act of 1969, 42 U.S.C. §4321 et seq.

National Research Council. (1989). *Improving risk communication.* Washington, DC: National Academy Press.

Palenchar, M. J. (2008). Risk communication and community right to know: A public relations obligation to inform. *Public Relations Journal, 2,* Article 0001a.

Palenchar, M.J. (2011). Concluding thoughts and challenges. *Management Communication Quarterly, 25*(3), 569–575.

Palenchar, M.J., & Heath, R.L. (2006). Responsible advocacy through strategic risk communication. In K. Fitzpatrick and C. Bronstein (Eds.), *Ethics in public relations: What is responsible advocacy?* (pp. 131–153). Thousand Oaks, CA: Sage.

Pellow, D.N., & Brulle, R.J. (2005). *Power, justice, and the environment: The critical appraisal of the Environmental Justice Movement.* Cambridge, MA: MIT Press.

Rabban, D.M. (1985). The ahistorical historian: Leonard Levy on freedom of expression in early American history. *Stanford Law Review, 37*(3), 795–856.

Rayner, S. (1992). Cultural theory and risk analysis. In S. Krimsky & D. Golding (Eds.), *Social theories of risk* (pp. 83–115). Westport, CT: Praeger.

Renn, O. (1992). Concepts of risk: A classification. In S. Krimsky & D. Golding (Eds.), *Social theories of risk* (pp. 53–79). Westport, CT: Praeger.

Resource Conservation and Recovery Act of 1976, 42 U.S.C. §6901 et seq.

Seeger, M.W. (2006). Best practices in crisis communication: An expert panel process. *Journal of Applied Communication, 34*(3), 232–244.

Skanavis, C.T., Koumouris, G. A., & Petreniti, V. (2005). Public participation mechanisms in environmental disasters. *Environmental Management, 35*(6), 821–837.

Sutton, J., Palen, L., & Shklovski, I. (2008). *Backchannels on the front lines: Emergent uses of social media in the 2007 Southern California wildfires.* Proceedings of the 5th International ISCRAM Conference, Washington, DC.

Taylor, D. (1997). American environmentalism: The role of race, class, and gender in shaping activism, 1820–1995. *Race, Gender, and Class, 5,* 15–62.

Toxic Substances Control Act of 1976 15 U.S.C. §2601 et seq.

Viscusi, W.K. (2000). Risk equity. *Journal of Legal Studies, 29,* 843–871.

Webler, T., & Tuler, S. (2006). Four perspectives on public participation process in environmental assessment and decision making: Combined results from 10 case studies. *The Policy Studies Journal, 34*(4), 699–722.

CHAPTER SIX

Public Response Before and After a Crisis: Appeals to Values and Outcomes for Environmental Attitudes

Janas Sinclair

Barbara Miller

This chapter focuses on a type of marketing campaign that centers on core societal values and directly associates them with environmental issues—typically in conflict with pro-environmental communication goals. Marketplace advocacy campaigns are sponsored by corporations and industry associations to improve or protect the market for their products, usually by appealing to commonly held societal values. As public and media interest in energy and the environment has increased, these campaigns have come to represent a significant portion of all advocacy campaigns. Although these campaigns may target specialized groups, oftentimes they are targeted toward the general public to build widespread support and counteract current or potential concerns about environmental, health, and/or societal risks associated with the product. This chapter presents an empirically tested model of audience response to marketplace advocacy campaigns and discusses the model in the context of ongoing proactive communication efforts as well as corporate crisis scenarios. Communication implications are explored.

Marketplace Advocacy

A prominent television advertisement initially aired during the 2008 Beijing Olympics features a crane taking flight along a beautiful Chinese coastline. Other cranes wait patiently in line for take-off as sea turtles (i.e., "ground traffic") cross the pristine beach. The narration prompts audiences to *"Imagine a way to fly that not only helps save millions of gallons of fuel, but actually reduces emissions."*

Another television advertisement using CG animation features the reconstruction of a Kansas town destroyed by a tornado, the lighting of a remote village by solar power, floating cars in parking lots, and a lush corn field. The narrator explains that *"Open science...can rebuild cities and make them more sustainable. It can bring solar power to remote villages. It makes materials lighter which saves fuel. It can feed a growing planet."*

The sponsors of these ads? General Electric and DuPont.

Their objective? Not to sell GE or DuPont products but rather to position these companies as environmentally responsible and concerned corporations. The first ad is one of several featured prominently in GE's "Ecomagination" campaign to bridge technology and the environment. The second ad is part of DuPont's "Open Science" campaign to promote efforts to make industries more eco-friendly and tap new energy sources.

Such campaigns epitomize marketplace advocacy, a form of corporate issue advocacy used to respond to burgeoning societal concerns, particularly regarding environmental issues. Marketplace advocacy uses marketing communication techniques to promote business' efforts while addressing public concerns toward the product itself or its physical and environmental impacts (Arens, 2004; Miller, 2010; Sinclair & Irani, 2005), usually by appealing to commonly held societal values (Bostdorff & Vibbert, 1994). Understanding how the public responds to marketplace advocacy, particularly environmental marketplace advocacy, is important given that an identified outcome of marketplace advocacy involves reducing the potential for future government intervention in corporate activities (e.g., Cutler & Muehling, 1989; Miller & Sinclair, 2009a; Miller & Sinclair, 2009b; Sethi, 1977; Sinclair & Irani, 2005). This outcome intensifies an already identified challenge for the environmental movement: lack of leadership for change among policy makers and lack of demand from the electorate for regulatory in-

tervention (Crompton & Kasser, 2009). Environmental organizations must understand how the public responds to marketplace
advocacy messages to develop effective counter campaigns. In
"Weathercocks and Signposts," Crompton (2008) argues that the
current "marketing" approach to pro-environmental campaigns is
flawed because it fails to engage core values and identities. He
calls for environmental campaigns to follow the lead of political, as
opposed to marketing, campaigns and focus on clearly articulating
the foundational values required to generate acceptance for the
more radical lifestyle and societal changes necessitated by today's
global environmental crises. While political campaigns may indeed
serve as a model, Crompton's analysis fails to also consider *marketers'* efforts at issue advocacy, namely marketplace advocacy,
which not only centers on core societal values but directly associates them with environmental issues.

This chapter answers Crompton's (2008) call for an examination of the ways in which marketers influence attitudes toward
environmental issues. This chapter focuses specifically on environmental marketplace advocacy that engages core societal values
and directly associates them with environmental issues—typically
in conflict with pro-environmental communication goals. The underlying message of corporate marketplace advocacy is that because environmental challenges can be solved through advances in
science and technology, adherence to current marketplace models
is acceptable (and advisable). In contrast, pro-environmental messages advocate for changes in consumer behavior and economic
models, ranging from calls for minor behavioral modifications to
the complete paradigm shifts espoused by Crompton (2008) and
other representatives of the environmental movement.

In the following section marketplace advocacy campaigns are
discussed as a tactic that engages audience members' core values
and identities—and are therefore relevant to efforts to either
change or maintain current environmental attitudes and behavior.
Environmental marketplace advocacy is defined, and the goals and
characteristics of these corporate messages are discussed. Next,
the chapter presents a model of audience response to corporate

marketplace advocacy including theoretical foundations and empirical evidence. This model is discussed in the typical advocacy scenario of building public support for a company and countering current or potential criticism as well as the less usual, but critical, scenario of corporate crisis. Finally, ethical implications are discussed and communication recommendations are presented.

The focus of this chapter is on marketplace advocacy, a type of issue advocacy that involves business' efforts to promote a product or service while addressing public concerns toward risks associated with the product or service or the manufacturing processes used to create it (Arens, 2004; Miller, 2012; Miller, 2010; Miller & Sinclair, 2009a; Miller & Sinclair, 2009b; Sinclair & Irani, 2005). As a general category, issue advocacy is considered distinct from other forms of marketing communication because it involves an issue, or topic, that could be considered controversial, rather than the more straightforward and typical goal of promoting a brand or the corporate image. Issue advertising may focus on marketplace, political, or values issues (Arens, 2004)—and it frequently combines aspects of all three. Marketplace advocacy campaigns often arise in response to burgeoning societal concerns, such as those faced by many industries, that may result in low credibility and trust among various segments of an organization's audience. In many cases, marketplace advocacy campaigns are initiated to communicate an organization's stance on an issue in an effort to sway public sentiment or generate support (Cutler & Muehling, 1989; Miller, 2010) and maintain a climate supportive of business activities (Gandy, 1982). This promotion of business interests may occur in a number of ways, including deflecting criticism, promoting an organization's image, laying the groundwork for future policy debates, and fostering the values of the free enterprise system (Bostdorff & Vibbert, 1994; Cutler & Muehling, 1989; Sethi, 1977).

"Greenwashing" has been used to describe marketplace advocacy ads that exaggerate the extent and effects of corporate environmental initiatives, and such overstated claims may foster the impression that voluntary corporate efforts make government regulation unnecessary. Even when corporate environmental efforts are voluntary—which is not always the case, as the environmental change may have been initiated under threat of governmental prosecution (Sethi, 1977)—the millions of dollars spent on the advocacy campaign to tout environmental achievements can overshadow the company's minor environmental improvements. According to Greenpeace (2010), for some businesses, efforts to promote an ecological conscious may be "little more than a convenient slogan....At best, such statements stretch the truth; at worst, they help conceal corporate behavior that is environmentally harmful by any standard" (Greenpeace, 2010).

Marketplace advocacy may be launched by a single sponsor, such as GE's "Ecomagination" and DuPont's "Open Science" campaigns, or sponsored by an industry association that represents the collective interests of the sector. For example, industry campaigns have emphasized the role of coal in meeting growing energy needs (American Coalition for Clean Coal Electricity: "America's Power") and crops produced through biotechnology (Council for Biotechnology: "Good Ideas Are Growing"). Marketplace advocacy campaigns can also be local in scope; the West Virginia Coal Association began the "Friends of Coal" campaign in West Virginia in 2002 to promote and gain support for the coal industry and public policies that affect the industry.

Appealing to Values

Just as political advocacy campaigns attempt to associate a candidate with values such as pride and/or hard work, marketplace advocacy emphasizes values to build a "reservoir of credibility" regarding future public policy issues (Bostdorff & Vibbert, 1994, p. 150). Marketplace advocacy may be a particularly effective means of both image building and influencing policy because of its unique ability to persuade without seeming to do so. Frequently incorporating a strong values advocacy component, its simplistic nature belies the potential of marketplace advocacy to distract attention from serious questions about organizational policies and public issues (Bostdorff & Vibbert, 1994). If environmentalists can be faulted for "putting the technical policy cart before the vision-and-values horse" (Shellenberger & Nordhaus, 2004, p. 23), market-place advocacy certainly avoids this problem, as these campaigns usually include relatively brief and selective references to corporate activities and a much stronger emphasis on commonly shared values. Message strategy often praises societal values, condemns oppositional values, discusses philanthropic efforts, and/or associates an organization's products with worthwhile societal goals.

In the 1970s, for example, Phillips Petroleum engaged in a campaign promoting the company's contribution to the public good, including the development of a blood filter for kidney patients and a fuel additive that helped make a helicopter rescue possible from a snowy mountain (Bostdorff & Vibbert, 1994). More recently, ads for the American Plastics Council's "Plastics Make It Possible" campaign have similarly highlighted how the plastics industry has made societal contributions, emphasizing the "everyday miracles" of plastics in child and food safety products. A common theme among many coal-related advocacy campaigns, meanwhile, involves an appeal to regional and national pride. The "Friends of Coal" campaign by the West Virginia Coal Association emphasizes "hard-working coal miners," as well as the industry's heritage and role in powering a growing nation (West Virginia Coal Association, 2009). On a national level, the "America's Power" campaign by the American Coalition for Clean Coal Electricity appeals to U.S. nationalism, encouraging website visitors to join "America's Power Army" and promoting "clean coal" as America's most abundant energy source (America's Power, 2010).

Another underlying theme throughout much marketplace advocacy is the value of determination and ingenuity as tools to overcome seemingly insurmountable challenges—with the implication that no drastic changes are needed in current business practices or consumption patterns. While rhetorical discourses vis-à-vis business and the environment may both be associated with traditional values and themes, these values are often in direct opposition (Cox, 2010), with corporate interests advocating for efficient use of natural resources for growth and economic development (Cox, 2010) and environmental advocates decoding the American experience as retaining a state of natural innocence (Brown & Crable, 1973)

Indeed, the GE and DuPont ads referenced in the opening of this chapter seem to promise that corporate research and development (or "open science") will solve problems related to energy, pollution, and sustainability through development of new, marketable products. Another GE ad in the "Ecomagination" campaign similarly promotes the value of ingenuity and the ability of scientific progress to overcome environmental problems by developing new products. This ad reflects the classic story, *The Little Engine That Could*: "Can technology and the environment peacefully coexist? Ecomagination answers yes with the Evolution Series locomotive....This is the little engine that could. And will." Similarly, in an ad featuring leafy green trees, computer-generated schematics of hydrogen cars, mathematical equations, and scientists working in laboratories, an Exxon Mobil engineer explains how the company is "working with partners to develop new energy saving technology for future decades" including a fuel cell car that "could enable about 80% better fuel economy than the car you and I drive today." The Exxon Mobil ad features the tagline "Taking on the World's Toughest Energy Challenges."

Marketplace Advocacy Spending

Shellenberger and Nordhaus (2004) might describe these ads as an appeal to the "fantasy of technical fixes" and the "siren call of denial" which, according to "The Death of Environmentalism," characterize public opinion on the environment (p. 5). The desire to overlook environmental problems may in fact be reinforced by marketplace advocacy, and environmental organizations must understand how the public responds to these messages to develop effective counter campaigns, particularly given marketplace advocacy's share of voice on environmental issues. GE, for example, spent $150 million just in 2006 on media buys for the "Ecomagination" campaign (Ad Age, 2007)—a figure unlikely to be matched by non-profit organizations, which typically rely on donations of ad time and space from media organizations to run their advertising messages.

Indeed, data reveal the dominance of corporate voices in advocacy campaigns. The Annenberg Public Policy Center's ongoing study involving print and television issue ads in the Washington, D.C. area found that campaigns focusing on 1) business/economy and 2) energy/environment were two of the top three issues covered in issue advertising during both the 107th and 108th Congress. Of the total spending on legislative issue advertisements during the 108th Congress ($404 million), 79% involved corporate interests rather than citizen-based/cause advocacy groups (Annenberg Public Policy Center, 2005). Additional content analyses of issue advertising in newspapers (Waltzer, 1988; Brown et al., 2001) and on-line (Williams, Foxman, & Saraswati, 2007) provide further evidence of the dominance of corporate-sponsored messages over those sponsored by citizen-based or other cause advocacy groups. The dominance of corporate sponsored advocacy messages, particularly in the realm of political advocacy, may continue to rise. The U.S. Supreme Court's 2010 decision in *Citizens United v. Federal Election Commission* relaxed restrictions on corporate political advocacy, including prior limitations on direct statements to vote for or against a candidate, which potentially gives corporations power to influence government policy.

Campaign Objectives

While in some cases marketplace advocacy may target specialized audiences, such as lawmakers, these messages are often aimed at the general public in an effort to build widespread support and counteract current or potential concerns about environmental, health, and/or societal risks associated with the product (Bostdorff & Vibbert, 1994; Sethi, 1977). Specific campaign objectives include counteracting perceived media bias and influencing public opinion, policy debate, and legislation. Marketplace advocacy campaigns are launched in response to current or anticipated controversy, and are therefore designed to build trust while deflecting criticism of the organization, its policies, products, or services (Bostdorff &

Vibbert, 1994). In that way, campaigns can indirectly reduce the potential for future government intervention in corporate activities that results from public calls for investigations of, or protection from, industry (Cutler & Muehling, 1989; Sethi, 1977). Marketplace advocacy campaigns may also lay the groundwork for counteracting future legal or policy arguments by increasing the potential for audience adherence to certain values that may contradict values that are likely to be invoked by the opposition. Generally, corporate sponsors may emphasize that environmental problems can be solved in the context of current models of production and consumption, while environmental groups may espouse new values and change. Campaigns have been used by American Electric Power to reduce regulations on coal mining, including strip mining and pollution codes; by Mobil Oil to prevent legislative passage of an excess profits tax directed at oil companies; by the Chrysler Corporation to slow the implementation of automotive pollution controls; and by Bethlehem Steel to restrict steel imports (Cutler & Muehling, 1989).

Organizations have also initiated marketplace advocacy campaigns in response to a perceived lack of objectivity by the media and an anti-business media climate (Fox, 1986; Miller, 2010). According to Cutler and Muehling (1989), "Extensive attacks by the media, unbalanced by recognition of business' contribution to the American standard of living, have contributed to what might be referred to as business' persecution complex" (pp. 41–42). Oil, gas, and other energy companies, in particular, have used marketplace advocacy campaigns to respond to media criticism on environmental and energy conservation issues. Many of these campaigns, however, downplay industry's adverse effects on the environment by exaggerating the often-miniscule efforts of industries to control pollution or improve their environmental record while publicizing adverse effects to the economy that may result from various environmental restrictions (Miller, 2010; Sethi, 1977), and thereby cross the nebulous line into greenwashing.

A Model of Public Response to Marketplace Advocacy

Though grassroots public relations and lobbying may be strategies within an overall campaign, marketplace advocacy frequently involves a *mass* media approach, as evidenced by GE's "Ecomagination" sponsorship of the 2006 Beijing Olympics. Marketplace ads for GE, Exxon Mobil, BP, America's Power, and DuPont, to name just a few, can be seen on television, in magazines, and in out-of-home venues such as airports. Both the values appeal and the message placement strategy are designed to target a widespread audience rather than narrowly defined groups. While certain segments of the population, namely those with high levels of environmental concern, interest, and/or expertise, may reject all or parts of these messages on face validity, how do lay audiences with little or no expertise on the environmental issue or related science make sense of these messages?

The Persuasion Knowledge Model (PKM) provides a useful guide for identifying the beliefs that people access in response to a persuasive situation (Friestad & Wright, 1994; Campbell & Kirmani, 2000; Kirmani & Campbell, 2004; Campbell & Kirmani, 2008). According to the PKM, three types of consumer knowledge affect the outcome of a persuasion attempt: 1) topic knowledge, 2) agent knowledge, and 3) persuasion knowledge. While topic knowledge is relevant to the informational focus of a message, the other types of knowledge pertain to the persuasion experience itself. Agent knowledge encompasses beliefs about the goals and characteristics of the message sponsor, and persuasion knowledge focuses on beliefs about the purpose of a specific persuasive tactic, an audience member's own goals in responding to the tactic, and the actions one can take to manage the persuasion attempt. According to Friestad and Wright (1994), many different goals drive responses to persuasion attempts. These goals include identity goals, such as managing other peoples' impressions or one's own self-image. Another coping goal is managing one's long-term relationship with the message sponsor, or agent. Consumers may attempt to manage this relationship, for example, by developing and maintaining accurate attitudes toward the sponsor so they can better assess its motives and future behavior.

We posit that two types of perceptions are central to audiences' responses to environmental marketplace advocacy: perceptions of the sponsor's accountability and the message's trustworthiness. Perceptions of sponsor accountability can be classified as both the salient agent *and* topic knowledge, because in the context of marketplace advocacy, the activities of the persuasive agent (the ad sponsor) are also the focal topic of the message. Perceptions of the message's trustworthiness represent the salient persuasion knowledge. As Campbell and Kirmani (2008) state, beliefs about the persuasion agent and the persuasion situation may include overlapping elements, but they are conceptualized as unique facets of knowledge, and research supports the unique contribution of these two types of perceptions on responses to marketplace advocacy. As shown in Figure 1, perceptions of sponsor accountability and message trustworthiness have been found to lead to persuasion coping outcomes that are favorable to the advertiser. More specifically, when perceptions of accountability and trustworthiness are *strong*, audience members are *more* likely to identify with the values in a marketplace advocacy message and *less* likely to critically evaluate the message. In contrast, when perceptions of accountability and trustworthiness are *weak*, audience members are *less* likely to identify with message values and *more* likely to engage in critical evaluation. These audience responses— identification with values and critical evaluation—impact the key persuasive outcomes of attitude towards the advertiser, with a positive impact for the former and a negative impact for the latter. The model is supported by findings from studies focusing on biotechnology (Sinclair & Irani, 2005) and coal advocacy campaigns (Miller, 2012; Miller & Sinclair, 2009a; Miller & Sinclair, 2009b). Audience members' existing level of environmental concern is also expected to have an impact on the attitudinal outcomes for environmental advocacy campaigns.

Figure 1. **Model of Public Response to Marketplace Advocacy**

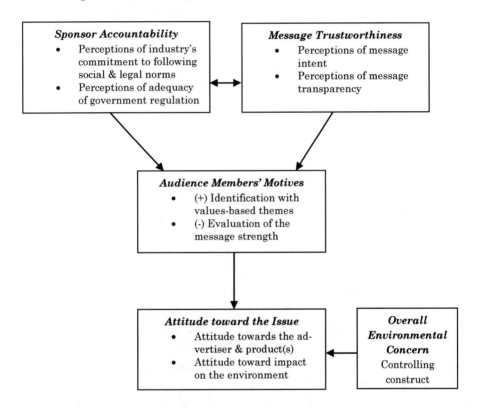

Sponsor Accountability

In the PKM, agent knowledge has been conceptualized as including both beliefs about an individual persuasive agent and generalized beliefs about the category of agents represented (Campbell & Kirmani, 2008). When marketplace advocacy is sponsored by an industry organization, such as the American Plastics Council or Council for Biotechnology Information, perceptions can be clearly defined at the category—or industry—level (Miller & Sinclair, 2009a; Miller & Sinclair, 2009b; Sinclair & Irani, 2005). Past research indicates perceptions of industry accountability may be salient when audiences form attitudes toward marketplace advocacy campaigns and that accountability perceptions include two dimensions: perceptions of the industry's commitment to following social and legal norms and perceptions of the adequacy of government regulation. In a qualitative study of attitudes towards a coal industry campaign, focus group participants discussed beliefs about the coal industry's commitment (or lack thereof) to community welfare, including the degree to which the industry held itself accountable to the community and the degree to which regulators held the industry accountable (Miller & Sinclair, 2009a). Quantitative data also support the unique influence of these dimensions on public response to marketplace advocacy (Miller & Sinclair, 2009b).

These findings are consistent with conceptualizations of interpersonal accountability, defined as an actor's answerability to audiences for fulfilling obligations, duties, and expectations (Schlenker et al., 1994; Schlenker & Weigold, 1989; Schlenker, Weigold, & Doherty, 1991). When individuals in a society are held accountable for their actions, citizens believe society's rules will be followed, and if the rules are broken, offenders will be appropriately sanctioned. Accountability was also examined in a study of consumer response to ads from the Council for Biotechnology Information's "Good Ideas Are Growing" campaign, where perceptions of industry concern and the adequacy of government regulation had a significant effect on attitudes toward ads for genetically modified crops (Sinclair & Irani, 2005).

Message Trustworthiness

Findings also support the influence of perceptions of message trustworthiness on responses to marketplace advocacy campaigns. Message trustworthiness has been found to include two salient dimensions: perceptions of the message intent and perceptions of message transparency. In focus group research evaluating attitudes toward a coal industry campaign (Miller & Sinclair, 2009a), participants questioned the intent of the messages (e.g., "What propaganda are they trying to shove down our throat now?") and the transparency (e.g., "What did you really do? What are you doing?"). Similarly, a quantitative test indicated the two dimensions of trust were predictive of responses to marketplace advocacy (Miller & Sinclair, 2009b). These findings are consistent with definitions of trust in the psychological and marketing literature as the expectation that promises will be kept and obligations fulfilled (Rotter, 1971; Barber, 1983; Dwyer, Schurr, & Oh, 1987). Trust has been identified as a key concern for advocacy advertising messages in the face of audience expectations of corporate self-interest (Fox, 1986), and perceptions of trust were found to predict attitudes toward the Council for Biotechnology Information's "Good Ideas Are Growing" campaign (Sinclair & Irani, 2005). Haley (1996) specifically found that perceptions of the intent dimension of trust were key to audience interpretation of health-related advocacy messages.

Audience Motives/Persuasion Coping

In the Persuasion Knowledge Model, Friestad and Wright (1994) outline a variety of coping strategies that audiences pursue in response to a persuasion attempt, including managing outcomes for their own identity and holding valid attitudes toward the persuasive agent to better understand it and manage their on-going interactions with the marketer. Qualitative (Miller & Sinclair, 2009a) and quantitative research (Miller & Sinclair, 2009b) indicates that these motives are relevant for audience response to advocacy campaigns. Focus group participants identified with the positive values in the ad concepts ("It reflects the pride that we have, and the miners have, in bringing that power to our country") and also frequently rejected message claims based on inconsistencies between the message and their own knowledge of the industry. Quantitative testing provides support for the impact of perceptions of accountability and trustworthiness on identification with the themes in a coal industry advocacy campaign and evaluation of the information content of the campaign (Miller & Sinclair, 2009b).

These findings for the effects of identification with the message and evaluation of the message are consistent with other research on persuasion and advocacy messages in particular. Perceptions that the message sponsor's values are congruent with one's own values have been identified as a key component of attitudes towards advocacy advertising (Haley, 1996). Research indicates that perceptions of trust and credibility lead to message acceptance (Newell & Goldsmith, 2001; MacKenzie & Lutz, 1989), which in the case of marketplace advocacy can be expected to take the form of identification with the values in the message. When the general public encounters marketplace advocacy messages, it seems relatively low involvement could be expected for many audience members across most situations. In the Elaboration Likelihood Model (ELM) framework, under low involvement, elements of source credibility are expected to serve as peripheral cues (Petty & Cacioppo, 1986). When perceptions of sponsor accountability and message trustworthiness are strong, therefore, marketplace advocacy campaigns employing a values-based persuasion strategy can be expected to induce an individual to focus on simple but compelling message elements (Perloff, 1993), enhancing identification with the positive message themes and reducing critical evaluation of the message. Thus, an information processing perspective supports the proposition that perceptions of industry accountability and sponsor trustworthiness may not only lead audiences to identify with the values presented in the ad but also to evaluate the message less critically.

Attitude Toward the Issue

Identification with the ad values coupled with reduced critical evaluation of the ad message (persuasion coping) influences overall attitude toward the advertiser and the issue, the key outcomes for marketplace advocacy campaigns. The Council for Biotechnology Information campaign study found perceptions of industry accountability and trust in the message sponsor were both related to attitudes toward the ad (Sinclair & Irani, 2005), and findings from the qualitative study of the coal industry campaign similarly support this relationship and also provide more specific insight into the processes by which trust and accountability impact attitudes (Miller & Sinclair, 2009a). Further, a quantitative test of the overall model of public response to marketplace advocacy accounted for approximately 80% of variance in attitudes toward the advocacy campaign (Miller & Sinclair, 2009b).

Public Response in Times of
Corporate Crisis and Non-Crisis

As mentioned above, the goal of marketplace advocacy campaigns is to build general support among the public and address potential concerns that may arise about a company (or industry) and its policies, products, or services, while also deflecting actual or potential criticism from minority segments. In some aspects, marketplace advocacy has objectives analogous with crisis communication. A crisis situation involves a major, unpredictable, low probability event that may have a potentially significant impact on an organization and its stakeholders (Massey, 2001; Miller & Horsley, 2009; Pearson & Clair, 1998). Crises tend to escalate in intensity, involve a short response time, fall under close media or government scrutiny, interfere with normal operations, jeopardize the positive public image of a company, and have the potential to damage the company's bottom line (Fink, 2002; Fishman, 1999). Following a crisis, in addition to managing all technical aspects of a crisis, crisis communication is employed to communicate with key stakeholders, rebuild a damaged reputation (White & Mazur, 1995), and minimize harm to an organization, including damage to reputation or lawsuits (Heath, 2006).

Although public and media attention on environmental issues, and relevant governmental action, can significantly impact a corporation's bottom line in much the same way a crisis situation might, environmental marketplace advocacy is part of a corporation's overall marketing efforts to influence public opinion in the hopes of building a "reservoir of credibility" regarding future public policy issues (Bostdorff & Vibbert, 1994, p. 150). Unlike crisis communication, however, marketplace advocacy campaigns are launched in times of non-crisis. Through ongoing, proactive communication, environmental marketplace advocacy encourages the public to focus their concerns on that which reflects favorably on the corporation (or industry) while minimizing that which might raise concern and possibly lead to governmental action. In a study of the agenda-building influence of a coal industry-sponsored marketplace advocacy campaign, for example, Miller (2010) found that marketplace advocacy influenced the public's agenda regarding industry-related issues. Not only were audiences well aware of industry-related issues promulgated by the campaign, such as energy and the economy, but it seems they were also accepting of industry-advanced benefits regarding these issues, particularly coal mining's heritage, the contribution of coal mining to the state and national economy, and the ability of coal to generate electricity for homes and modern conveniences. Moreover, this agenda-building influence among stakeholders resulted in more favorable attitudes toward the overall industry.

Response to Marketplace Advocacy in Times of Corporate Crisis

Consumer response to corporate marketplace advocacy campaigns, of course, is expected to be influenced by a corporate crisis. Research has not yet examined how the model of public response to marketplace advocacy operates in times of crisis, but perceptions regarding sponsor accountability and message trustworthiness are likely to be particularly salient in this scenario. While corporations and industry groups typically suspend their usual marketplace advocacy campaigns during a crisis in lieu of tailored crisis communication, for any type of corporate communication encountered in the midst of a crisis, motives to critique the message could be especially dominant relative to motives to identify with positive message themes. Accountability issues may be particularly salient in terms of perceptions of the industry's commitment to following social and legal norms as well perceptions of the adequacy of current government oversight. Similarly, audiences are likely to be particularly skeptical of the intent and transparency of the message, and therefore less likely to find it trustworthy.

A related issue is the effect that prior exposure to a marketplace advocacy campaign may have on public response to the corporate crisis itself and the company's crisis communication. While the goal of marketplace advocacy is to build favorable public opinion that will help the company to weather or avoid future criticism, in the event of a major crisis, prior marketplace advocacy could potentially lead to more *negative* public response, particularly if the crisis indicates inconsistency between the company's actions and its marketplace advocacy positioning as an environmentally responsible organization. Such inconsistency could prompt thoughts about lack of industry accountability, lack of trust in the message, and a critical evaluation of the message. The 2010 BP Deepwater Horizon oil spill in the Gulf of Mexico, for example, has been called the worst environmental crisis in U.S. history, resulting in an audience that is "livid at BP, angry at the Federal response and so sick of 'spin' that @BPGlobalPR, a satirical Twitter feed, now has over 120,000 followers" (Bush, 2010). While the extent of this disaster and the response by BP and the government could clearly elicit public outrage in itself, the negative public response may have been magnified by BP's prior marketplace advocacy campaign in which it was positioned as the "environmentally gentle oil company" (Garfield, 2010). In fact, a similar argument was made in 2006 after an arguably less severe crisis for BP when its Prudhoe Bay pipeline was found to be corroded and leaking. *New York Times* writer Joe Nocera wrote about a BP ad he encountered in an airport after the crisis broke:

> It spoke to the company's commitment to the environment. And here's what I thought when I saw it: "Yeah, right.."...If BP hadn't been so holier than thou in its marketing these past years, I doubt that it would be getting hammered right now—at least to this extent. (Nocera, 2006)

While crisis may serve to increase critical response among many audience members, it is also possible that anxiety about the environment could be heightened at this time, and in response some audience members may avoid thinking deeply about the issues (Crompton & Kasser, 2009) and instead be motivated to identify with the comforting, positive themes likely to be present in corporate crisis communication or marketplace advocacy messages. Individual difference variables could be expected to moderate the saliency of motives to critique or identify with corporate messages during crisis. For example, higher levels of environmental concern, need for cognition, and personal knowledge of current events related to the crisis could result in enhanced motives to critique the message despite fear and anxiety related to the crisis.

Conclusion

Consideration of marketplace advocacy and public response to corporate messages raise implications for companies, marketing communication practitioners, and organizations seeking pro-environmental change through their own communication campaigns. Research on public response to marketplace advocacy indicates that audiences respond to these messages based on perceptions of the accountability of the message sponsor and the trustworthiness of the messages. Corporations, therefore, must establish that they are committed to protecting the public interest by following social and legal norms for conduct. Perceptions of accountability can be further enhanced by referring to government oversight and, ideally, indicating how the company exceeds minimal requirements. Similarly, messages should be transparent and communicate a positive, rather than primarily self-interested, intent. A corporate crisis scenario highlights the importance of corporations not just claiming they are environmentally friendly but establishing accountability through their actions. As *New York Times* writer Joe Nocera stated, "If there is one ironclad rule about marketing, it is that you had better be practicing internally what you are preaching to the world. Let me put it another way: You can't just talk the talk, you have to walk the walk" (2006). *Advertising Age's* Rance Crain provided further advice to communication practitioners in light of the BP crisis, "If you are the ad agency, you'd better do some due diligence to make sure the glorified image you are creating for your client holds up when your client turns out to be not quite so enlightened as you depicted it to be" (Crain, 2006).

For pro-environmental communication, consideration of marketplace advocacy raises both opportunities and challenges. The dominance of corporate advocacy messages over those from nonprofit and citizen-based organizations due to extensive media budgets certainly presents a challenge. Moreover, the content of marketplace advocacy messages poses a significant challenge, with messages that directly conflict with the argument that major lifestyle and societal changes are required to address current environmental problems. On the other hand, examining public response to these corporate messages provides insight into the role of identity motives in response to environmental communication. Findings indicate that people are motivated to identify with positive imagery and solutions to problems, and it seems pro-environmental communication could similarly profit from positive imagery associated with nature, being a responsible part of the natural community, and taking action to help solve problems.

This chapter also provides a basis for the goal of counteracting the influence of advertising through education, which is outlined in "Meeting Environmental Challenges: The Role of Human Identity" (Crompton & Kasser, 2009). Crompton and Kasser (2009) point to advertising's general contribution to materialism—and negative impact on environmental concern—and call for media literacy programs to help people recognize persuasive techniques. Research on public response to marketplace advocacy provides insight into the specific types of information that might heighten critical response to the ad messages that are potentially most threatening to the environmental cause: those that communicate no real change is necessary. Directly addressing issues related to industry accountability and message trustworthiness in pro-environmental communication could be a particularly effective approach in the face of marketplace advocacy campaigns. For example, education campaigns could provide information that would aid the public in evaluating the true scope of industry commitment to environmental efforts, the amount of government oversight (or lack thereof), the intent of marketplace advocacy messages, and the transparency (or lack of transparency) of these messages. Corporate voices are already engaged in using values-based messages and identity appeals to advocate for the status quo in models of business and consumption. Examination of their strategies as represented in marketplace advocacy campaigns, and understanding of how the public responds to these messages, can aid in effectively presenting alternative viewpoints and advocating for the changes needed to meet the environmental challenges of today and the future.

References

Ad Age (2007). *100 leading national advertisers; top 200 brands.* Retrieved November 17, 2007, from http://adage.com/datacenter/#100_leading_national_advertisers_global_marketers_other_ad_spending_data.

America's Power (2010). Retrieved May 26, 2010, from http://www. americaspower.org/The-Facts/Power-House.

Annenberg Public Policy Institute (2005). *Legislative issue advertising in the 108th Congress.* Retrieved October 30, 2005, from http://www. annenbergpublicpolicycenter.org/issueads05/2003-2004/Source%20Files/APPC_IssueAds108thMM.pdf.

Arens, W. F. (2004). *Contemporary advertising* (9th ed.). Boston: McGraw-Hill.

Barber, B. (1983). *The logic and limits of trust.* New Brunswick, NJ: Rutgers University Press.

Bostdorff, D. M., & Vibbert, S. L. (1994). Values advocacy: Enhancing organizational images, deflecting public criticism, and grounding future arguments. *Public Relations Review, 20* (2), 141–158.

Brown, C., Waltzer, H., & Waltzer, M. B. (2001). Daring to be heard: Advertorials by organized interests on the op-ed pages of *The New York Times,* 1985–1998. *Political Communication, 18*(1), 23–50.

Brown, W. R., & Crable, R. E. (1973). Industry, mass magazines, and the ecology issue. *The Quarterly Journal of Speech, 59,* 259–272.

Bush, M. (June 7, 2010). Brunswick put to ultimate test as BP grows increasingly toxic. *Advertising Age, 81*(23), 2 & 37.

Campbell, M. C., & Kirmani, A. (2000). Consumers' use of persuasion knowledge: The effects of accessibility and cognitive capacity on perceptions of an influence agent. *Journal of Consumer Research, 27,* 69–83.

Campbell, M. C., & Kirmani, A. (2008). I know what you're doing and why you're doing it: The use of the Persuasion Knowledge Model in consumer research. In Curtis P. Haugtvedt, Paul M. Herr & Frank R. Kardes (Eds.), *Handbook of consumer psychology* (pp. 549–572). New York, NY: Taylor & Francis Group.

Cox, R. (2010). *Environmental communication and the public sphere* (2nd ed). Thousand Oaks, CA: Sage Publications.

Crompton, T. (2008). Weathercocks and signposts: The environmental movement at a crossroads. Retrieved from http://www.valuesandframes.org/downloads.

Crain, R. (2006). BP Should have concentrated on being a better oil company. *Advertising Age.* Retrieved from http://adage.com/article/rance-crain/bp-concentrated-a-oil-company/111406/

Crompton, T., & Kasser, T. (2009). Meeting environmental challenges: The role of human identity. Retrieved May 26, 2010 from http://www.wwf.org.uk/wwf_articles.cfm?unewsid=3105.

Cutler, B.D., & Muehling, D.D. (1989). Advocacy advertising and the boundaries of commercial speech. *Journal of Advertising, 18*(3), 40–50.

Dwyer, R., Schurr, P., & Oh, S. (1987). Developing buyer-seller relationships. *Journal of Marketing, 51*, 11–27.

Fink, S. (2002). *Crisis management: Planning for the inevitable.* Lincoln, NE: iUniverse.

Fishman, D. A. (1999). ValuJet flight 592: Crisis communication theory blended and extended. *Communication Quarterly, 47*(4), 345–376.

Fox, K. (1986). The measurement of issue/advocacy advertising effects. *Current Issues and Research in Advertising, 9*(1), 62–92.

Friestad, M., & Wright, P. (1994). The persuasion knowledge model: How people cope with persuasion attempts. *Journal of Consumer Research, 21*, 1–31.

Gandy, O. H. (1982). Public relations and public policy: The structuration of dominance in the information age. In Elizabeth L. Toth & Robert L. Heath (Eds.), *Rhetorical and critical approaches to public relations* (pp. 131-163). Hillsdale, NJ: Erlbaum.

Garfield, B. (June 21, 2010). From greenwashing to godwashing, BP and Obama fail at image control. *Advertising Age, 81*(25), 8.

Greenpeace (2010). Introduction to stop greenwash. Retrieved May 26, 2010, from http://www.stopgreenwash.org/introduction.

Haley, E. (1996). Exploring the construct of organization as source: consumers' understandings of organizational sponsorship of advocacy advertising. *Journal of Advertising, 25*(2), 19–35.

Heath, R. L. (2006). Best practices in crisis communication: Evolution of practice through research. *Journal of Applied Communication Research, 34*(3), 245–248.

Kirmani, A., & Campbell, M. C. (2004). Goal seeker and persuasion sentry: How consumer targets respond to interpersonal marketing persuasion. *Journal of Consumer Research, 31*, 573–582.

MacKenzie, S. B., & Lutz, R. J. (1989). An empirical examination of the structural antecedents of attitude toward the ad in an advertising pretesting context. *Journal of Marketing, 53*(2), 48–62.

Massey, J. E. (2001). Managing organizational legitimacy. *Journal of Business Communication, 38*(2), 153–182.

Miller, B. M. (2012). *Generating public support for business and industry: Marketplace advocacy campaigns.* Amherst, NY: Cambria.

Miller, B. M. (2010). Community stakeholders and marketplace advocacy: A model of advocacy, agenda building, and industry approval. *Journal of Public Relations Research, 22*(1), 85–112.

Miller, B. M., & Horsley J. S. (2009). Digging deeper: Crisis management in the coal industry. *Journal of Applied Communications Research, 37*(3), 298–316.

Miller, B. M., & Sinclair, J. (2009a). Community stakeholder responses to advocacy advertising: Trust, accountability, and the persuasion knowledge model. *Journal of Advertising, 38*(2), 37–51.

Miller, B. M., & Sinclair, J. (2009b). A model of public response to marketplace advocacy. *Journalism and Mass Communication Quarterly, 86*(3), 613–629.

Newell, S. J., & Goldsmith, R. E. (2001). The development of a scale to measure perceived corporate credibility. *Journal of Business Research, 52*(3), 235–247.

Nocera, J. (August 12, 2006). Green logo, but BP Is old oil. *New York Times,* p. C1(L).

Pearson, C. M., & Clair, J. A. (1998). Reframing crisis management. *Academy of Management Review, 23*(1), 59–77.

Perloff, R. M. (1993). *The dynamics of persuasion.* Hillsdale, NJ: Lawrence Erlbaum Associates.

Petty, R. E., & Cacioppo, J. T. (1986). The elaboration likelihood model of persuasion. In Leonard Berkowitz (Ed.), *Advances in experimental social psychology,* Vol. 19 (pp. 123–205). New York: Academic Press.

Rotter, J. (1971). Generalized expectancies for interpersonal trust. *American Psychologist, 26,* 443–452.

Schlenker, B. R., Britt, T. W., Pennington, J., Murphy, R., & Doherty, K. (1994). The triangle model of responsibility. *Psychological Review, 101*(4), 632–652.

Schlenker, B. R., & Weigold, M. F. (1989). Self-Identification and accountability. In Robert Giacalone & Paul Rosefeld (Eds.), *Impression management in the organization* (pp. 21–43). Hillsdale, NJ: Lawrence Erlbaum Associates.

Schlenker, B. R., Weigold, M. F., & Doherty, K. (1991). Coping with accountability: Self-identification and evaluative reckonings. In C. R. Snyder & Donelson R. Forsyth (Eds.), *Handbook of social and clinical psychology: The health perspective* (pp. 96–115). New York: Pergamon.

Sethi, S. P. (1977). *Advocacy advertising and large corporations.* Lexington, MA: Lexington Books.

Sinclair, J., & Irani, T. (2005). Advocacy advertising for biotechnology: The effect of public accountability on corporate trust and attitude toward the ad. *Journal of Advertising, 34*(3), 59–73.

Shellenberger, M., & Nordhaus, T. (2004). *The death of environmentalism: Global warming politics in a post-environmental world.* Retrieved May 26, 2010, from http://www.thebreakthrough.org/PDF/Death_of_Environmentalism.pdf.

Waltzer, H. (1988). Corporate advocacy advertising and political influence. *Public Relations Review, 14* (Spring), 41–55.

West Virginia Coal Association (2009). *Coal Facts 2009.* Retrieved May 26, 2010, from http://www.wvcoal.com/docs/Coal%20Facts%202009.pdf.

Williams, C. B., Foxman, E.R., & Saraswati, S. P. (2007). A comparative study of non-profit and for-profit web-based advocacy. *Journal of Political Marketing, 16*(1), 69–97.

White, J., & Mazur, L. (1995). *Strategic communications management: Making public relations work.* Wokingham, England: Addison-Wesley.

CHAPTER SEVEN

Individual Factors and Green Message Reception: Framing, Lifestyles and Environmental Choices

Harsha Gangadharbatla

Kim B. Sheehan

Interest in environmental communication is at an all-time high and the debate over the best strategies to bring about changes in pro-environmental behaviors is raging. Several environmentalists question the effectiveness of a foot-in-the-door approach (also called a marketing approach) to saving our planet. The current chapter begins by listing the advantages and disadvantages of a marketing approach and argues that a marketing approach may not be completely useless in achieving the goals of the environmental movement. More precisely, an experiment was conducted with Message Framing and the Lifestyle of Health and Sustainability (LOHAS) scale. Lifestyles of individuals as the two independent variables and their effect on individuals' intentions to engage in pro-environmental behavior is investigated. Results suggest that individuals who were exposed to an environmental message were more likely to recycle and conserve energy than those who were exposed to a non-environmental message. Further, individuals rating high on the LOHAS scale were more persuaded by the environmental ads than those low on the LOHAS scale.

In the ongoing debate over the environment and global warming, there is one side that believes that global warming is neither real nor man-made and another side that is increasingly alarmed about the future of our planet and our materialistic way of life. The percentage of Americans who believe that global warming is not real is estimated to be about 48% (Gallup, 2010). These Americans believe that the threat of global warming is exaggerated and is not due to human activities but it is due to natural causes. While the number of Americans who do not believe in global warming is on the decline, it is still surprising to some that the U.S. is one of the few developed nations where mainstream figures debate on TV about the legitimacy of global climate change (E.G., 2011). Of the people who do believe in global climate change, about 80% of them report buying "green" products (Smith, 2010), and 68% of them say they have taken steps in the past year to "green their lives," for instance, by improving the energy efficiency of their homes (The Nature Conservancy, 2010). Some academics, public policy makers, scientists, and researchers have gone further to suggest that a "foot-in-the-door" approach to saving our planet would not work and have instead called for more immediate and large-scale actions initiated and driven by governmental policy and regulation (Amidon, 2005; Crompton, 2008; Michaelson, 1998; Mowery, Nelson, & Martin, 2010). For example Mowery et al., (2010) recommends technological innovations and policy changes and regulations that can be employed across a wide variety of sectors on a global scale. Irrespective of where one falls on this spectrum of beliefs on global climate change and the possible solutions for it, the issue of global warming is not expected to fade away from mainstream media or our collective attention anytime soon (Kaufman, 2007). Both the issue of global climate change and the debate over what would be the best course of action to combat it are still highly contentious.

There are many reasons why Americans are reluctant to acknowledge global climate change as a real and man-made phenomenon with serious consequences that threaten our very existence. In *The Economist* article titled, "Why Don't Americans Believe in Global Warming?" Economist Group (2011) lists five broad reasons: psychological, economic, political, epistemological, and meta-physical. More precisely, the consequences of global warming are too grave for most humans to elaborate, so a denial of the issue rejects its consequences. These might be that costs of fighting global warming are too high, or we believe that any action in that area would drastically hamper our economic development and jobs; the issue of global warming is a political one rather than a scientific one; and finally, the debate over the validity of scientific evidence and faith in a higher power who will ultimately decide our fate (Economist Group, 2011). No matter what category a non-believer's arguments fall under, it is important for people who identify themselves as "environmentalists," defined as participants in a broad philosophy and social movement regarding concerns for environmental conservation and improvement of the state of the environment (Lincoln, 2009), to understand that public perceptions, beliefs, attitudes, and behavior are just as important as policy changes and regulation in the fight to reverse global warming.

Changing public perceptions, attitudes, and beliefs that lead to individuals' adoption of "simple steps" towards a sustainable life is referred to as a "foot-in-the-door" or "a marketing" approach to the environmental issue and is considered by many as ineffective and as an outright "distraction from the approaches that will be needed to create more systematic change" (Crompton, 2008, p. 6). In the following section, we will outline the advantages and disadvantages of a marketing approach to behavioral changes and argue that for environmental activists seeking to affect long-lasting change, the issue of public perceptions, attitudes, and behavior is just as important as the policy changes and governmental regulations that are recommended by Crompton (2008) and others to curb global warming.

The Good and Bad of the Marketing Approach

In "Weathercocks and Signposts," Crompton (2008) outlines the numerous disadvantages and the ultimate futility of a marketing approach to pro-environmental behavioral change. Crompton (2008) argues that the marketing approach is fundamentally flawed in that it relies on "simple and painless steps" that distract the environmental movement from a more systematic change that is needed. He contends that "emphasis on opportunities offered by 'green consumption' distract attention from the fundamental problems inherent to consumerism" (p. 6). According to him, appealing to the values that motivate individuals to engage in and activate public demand for governmental intervention might be the only surviving hope for humankind (Crompton, 2008).

While it is hard to argue with the idea that big and fundamental changes are needed to address this impending crisis, we are skeptical of the effectiveness of a forced or regulatory approach to pro-environmental behavior changes. We contend that changes in behavior and lifestyles (and ultimately, consumerism) are more effective and long-lasting when they occur in a bottom-up fashion rather than a top-down approach. And any such change would ensue only after an increased awareness of the environmental issue and acceptance that it is man-made. Given that almost half of Americans are still skeptical about global warming (Gallup, 2010), it is imperative that environmental activists work toward gathering more support for the cause rather than force governmental regulations on an unwilling and an unenthusiastic population.

Global advertising spending is estimated to be around $500 billion of which almost half is spent in the United States (Boris, 2011). The persuasion industry would not be spending such an exorbitant amount of money if advertising was not effective in influencing perceptions, attitudes, and behavior. Crompton (2008) argues that when it comes to environment, a marketing approach to behavioral changes is ineffective and distracts attention from the root of the problem, consumerism. It can be argued that the rampant consumerism, which environmentalists such as Crompton (2008) attribute our current environmental crisis to, is actually a product of the "marketing approach" that they are critical of. If the "marketing approach" can appeal to individuals and persuade the majority of us to consume, we argue that it can also do the same for the environmental movement. The "marketing approach" can persuade people to acknowledge that global warming is real and that the effects are grave and consequential if nothing is done.

We agree with Crompton (2008) in that environmental messages need to go beyond appeals to undertake "simple and painless" pro-environmental choices and take individuals' values and lifestyles into account. However, we do see the value of "simple and painless" pro-environmental changes in behavior such as conserving energy and recycling in building a long-term systematic change in our attitudes, beliefs, and behavior. If the environmental movement aims to garner support, gain legitimacy and enter the mainstream psyche, it needs to adopt the "marketing approach" along with other approaches.

The movement has been gaining momentum with several people taking steps to "green their lives" (Davis, 1995). For instance, there has been about 100% increase in the total recycling in the United States in the last 10 years but the U.S. only recycles about 32% of its waste (benefits-of-recycling.com, 2010). Numerous polls measure the number of people who are recycling or adopting other types of green behaviors, but more research is needed to understand why some people make these pro-environmental choices while some do not. In other words, what types of messages might be influencing individuals to make pro-environmental choices and how do these messages interact with their values and lifestyles? Even Crompton (2008, p. 6) acknowledges that the "values underpinning environmental behaviour will be of critical importance both in motivating individuals to engage in pro-environmental behavior...and in activating public demand for government intervention."

Changing public perceptions and behavior may be the first step toward reversing or curbing global climate change and for this to happen, it is important to understand what types of appeals (if any) work best in persuading what types of individuals. Framing the environmental issue in the most appealing way may yield more desirable results in terms of changing public opinions, beliefs, and behavior than forcing them with restrictive policies. Environmentalists need to understand the most effective way to frame this issue so as to not alienate non-believers but to also persuade believers to act.

The two ways in which issues are usually framed in persuasive communication are categorized as loss frame and gain frame. Loss frame deals with bringing to attention the dangers of not taking action whereas gain frame focuses on the benefits of taking action. Research has shown that individuals respond differently to different frames, particularly when it comes to environmental messages (Davis, 1995). Furthermore, not all messages appeal to all individuals equally. In the environmental area, individuals are different in terms of lifestyles and how they perceive different issues relating to health and sustainability. Individuals focused more on health and sustainabilty are often termed Lifestyle of Health and Sustainability or LOHAS consumers (LOHAS, 2008).

In the following section, we examine literature on framing and individuals' lifestyles and their role in persuasion. More precisely, we examine the interaction and direct influence of both message framing and the type of individual on the LOHAS scale when persuading individuals to engage in pro-environmental behavior, namely, recycling and conserving more energy.

Framing, Lifestyles, and Behavior

Framing an issue is often described as focusing attention on key elements of a message (Duhè & Zoch, 1994) by "selecting some aspects of perceived reality and making them more salient in the communicating text, in such a way as to promote a particular problem definition, causal interpretation, moral evaluation and/or treatment recommendation for [the particular issue]" (Entman 1993, p. 55). Framing is commonly used in persuasive communication such as advertising and public relations as individuals latch on to these "selective aspects" to connect to their underlying psychological processes that help them take in information, make evaluative judgments, and ultimately, draw inferences about a given issue (Hallahan, 1999). Framing is a key aspect of consumer socialization theory, which explains how consumers become consumers. By providing the proper frames, consumers chose specific lifestyles that they adopt for much of their adult lives (Moschis, 1987).

The most commonly researched frames in a wide variety of fields such as psychology, organizational decision making, economics, health communication, and political communication are loss and gain frames. The study of framing can be traced to the seminal work of Tversky and Kahneman (1981) that led to prospect theory. According to prospect theory, people respond differently to messages depending on how the messages are framed. A loss frame is when the message highlights the dangers or drawbacks associated with not taking a particular action, and a gain frame highlights the benefits or advantages associated with taking action over other things (De Dreu & McCusker, 1997; De Martino, Kumaran, Seymour, & Dolan, 2006; Tversky & Kahneman, 1981). The effect of loss versus gain framing has been studied extensively in other fields, but when it comes to environmental messaging, the research and evidence are limited.

We use Davis' (1995) study of framing the environmental problem in terms of loss versus gain for a target audience of current and future generations as our starting point. Davis (1995) examined the effect of framing the environmental problem for future and current generations in terms of actions, taking less and giving more, in a 2x2x2 design. His stimuli included eight short paragraphs written to represent the eight cells in the experimental design with attitude toward message, believability, relevance, importance, manipulativeness, and behavioral intentions as dependent measures. It was found that individuals do respond to environmental messages, and study participants were most influenced by "a communication which emphasized the negative consequences of their own inaction on themselves and their own generation" (Davis, 1995 p. 295). Overall, Davis' (1995) study establishes that individuals respond to simple, clear, and understandable messages that stress how the target will be directly and negatively affected if they fail to act.

In extending Davis' (1995) study we examine the role of a different target audience based on their rating on the LOHAS scale. In other words, we examine the role of loss versus gain framing in environmental communication but with a different type of classification of target audience. Instead of including the target audience in communication material, we obtained a mean score for each participant on the LOHAS scale that suggests whether the individual was high or low in terms of health and sustainability lifestyles. LOHAS describes an estimated $209 billion market with approximately 41 million people currently considered LOHAS consumers, which is about 19% of all adults in the United States (LOHAS, 2008). LOHAS-aligned individuals integrate healthier and more sustainable alternatives into every aspect of their lives by making consumption decisions that encourage sustainability in the areas of food and nutrition, home life, building and energy, and work and money (Höfer, 2009). With the LOHAS-aligned consumer market growing rapidly it becomes important to better understand how (if in any way) these individuals respond to different message frames compared to individuals that rate low on the LOHAS scale. Since individuals rating high on the LOHAS scale are already inclined to lead a healthy and sustainable lifestyle, our advertisements that address recycling should have greater influence on them than on individuals low on the LOHAS scale.

Based on the above discussion on message framing and LOHAS segmentation, we ask the following research questions:

RQ1: What impact (if any) do environmental messages have on individuals' intentions to recycle and conserve energy?

If environmental messages do indeed influence individuals, then our next question relates to the nature of this effect. As discussed in the earlier sections, individuals' predispositions towards environment and global warming are not all the same. Depending on their perceptions, beliefs, and attitudes towards various issues relating to health and sustainability, their opinions of this issue also vary. Environmentalists are often interested in changing the perceptions of non-believers. If targeting a particular group of non-believers is the objective, then it becomes imperative that one understand the nature of the relationship between message framing and the type of individual who is being targeted and how this impacts individuals' pro-environmental behavior.

RQ2: What type of appeal (loss versus gain) works best for what type of individual (based on their rating on the LOHAS scale)?

Method

In order to answer our research questions, we conducted an online experiment. Subjects were chosen from introductory courses at a large university in the northwest. An email was sent inviting them to participate in an online study. They were reminded both in the email and on the main page of the study that participation was purely voluntary and that they could stop and not be a part of the study at any time they wanted.

The stimuli was designed using a photoediting software, and the survey was created and hosted on a third-party web site called Surveymonkey. Subjects were randomly assigned to one of the three experimental conditions: a loss frame condition, a gain frame condition, and a control condition. Upon beginning the survey, the subjects responded to the LOHAS scale that included 10 five-point Likert scale items such as "I care about protecting the environment," "I care about sustainable agricultural practices," and "I care about using renewable energy sources." Subjects were next asked if they believed in global warming or not. The response was a simple yes or no. Additional variables measured included individuals recyling and energy conserving habits that were 10-point scale items with 1 being not likely at all and 10 being very likely to either recycle or conserve energy. There were a total of 12 such questions.

After responding to these questions, subjects were exposed to a full-page advertisement promoting recycling; each advertisement had the same dominant visual of a can. Recycling was selected since it is a fairly easy activity to adopt, yet only half of all paper in the US is recycled and the US was only given an 'average' score for its recycling efforts (Williams, 2008). There also appear to be no age differences in recycling behaviors. One group was exposed to an ad which appealed to the dangers of not taking action to recycle (or the loss frame) and the second was exposed to an ad that underscored the benefits of taking action to recycle (or the gain frame). The third group served as a control group where indviduals were exposed to a non-environmental message of a Sprite can. Each subject was exposed to only one ad.

Three ads were designed to represent the three conditions. As mentioned earlier, all three ads employed the exact same layout and design. They all included a close-up shot of a soda can with three different headlines. For the loss frame condition, the headline read, "Not recycling makes you an accomplice" with "I can ruin an eco system" in bold type on the soda can. For the gain frame condition, the ad's headline read, "The recycling industry employs over 1.1 million people" with "I can pay your bills" on the soda can. The third ad featured a close-up shot, again the exact same design and layout as the other two ads, of a Sprite can with the headline, "Can you taste the difference?"

Following the exposure to the advertisement, subjects were asked 12 questions about the credibility, relevance, the type of appeal, and persuasiveness of the ad. Finally, the questionnaire ended with demographic questions on age, gender, and ethnicity.

Results

A total of 329 subjects participated in the experiment of which 43% were men and 55% were women with the rest choosing not to respond to the gender question. The largest portion (75%) were white, followed by Asian (9%) and Hispanic (4%). Reliability assessment was conducted on each of the scales using Cronbach's Alpha with all exceeding the accepted level of .70. The mean scores, standard deviations along with the reliability indices are summarized in Table 1.

Table 1. Reliability, mean scores, and standard deviations of scales

Variable	Mean*	SD	α
LOHAS score (10 items)	3.67	.29	0.93
Sustainable behavior (12 items)	7.06	1.16	0.72

The LOHAS scale was then dichotomized at an approximate median score which resulted in 161 subjects rating low on the LOHAS scale and the other 168 rating high on it. This was done to obtain two levels of our second independent variable, yielding a 3x2 design where the first independent variable was exposure of a full-page ad with three levels—loss frame, gain frame, and control. Table 2 below shows the number of subjects in each cell with column and row totals.

Table 2. Number of subjects in each cell of our 3x2 design

	Loss Framing	Gain Framing	Control Group	Total
LOHAS—high	63	57	46	166
LOHAS—low	43	59	55	157
Total	106	116	101	323

Research Questions

In order to answer the first research question of whether environmental messages have any impact on individuals' behavior that is environment friendly, a simple one-way ANOVA was conducted (see Table 3) with exposure to environmental messages as an independent variable. Subjects saw ads from one of the three conditions: environmental ad with loss frame, environmental ad with gain frame, and a non-environmental ad for Sprite. The dependent variables measured their intentions to (1) recycle and (2) conserve energy after exposure to the ads.

Table 3. One-way ANOVA with "exposure to environmental messages" as IV

		Sum of Squares	df	Mean Squares	F-ratio	p-level
Intentions to recycle more	Between groups	82.22	2	41.16	40.55	<.0001
	Within groups	322.845	318	1.05		
	Total	405.184	320			
Intentions to conserve energy	Between groups	40.792	2	20.39	19.56	<.0001
	Within groups	331.56	318	1.04		
	Total	372.355	320			

Both F-ratios are statistically significant, suggesting that at least two of the means in between group comparisons are significantly different from each other. To examine which of these means of the dependent variables are different, we ran a post-hoc test, Tukey HSD.

Table 4a. Mean differences between groups; DV: intentions to recycle

Dependent Variable	Treatment (I)	Treatment (J)	Mean Difference (I-J)	Std. error	p-level
Intentions to recycle more	Loss frame	Gain frame	-.14	.13	.53
		Control	1.01*	.14	<.0001
	Gain frame	Loss frame	.14	.13	.53
		Control	1.15*	.13	<.0001
	Control	Loss frame	-1.01*	.14	<.0001
		Gain frame	-1.15*	.13	<.0001

* Mean Difference is significant at 0.05 level

Tables 4a and 4b show that individuals reported greater intentions to recycle and conserve more energy when exposed to environmental messages (gain or loss) than when exposed to a Sprite ad, which suggests that environmental messages do indeed activate the desire to be environmentally friendly, and this is irrespective of how individuals feel about global climate change or where they fall on the LOHAS segmentation scheme.

Table 4b. Mean differences between groups; DV: intentions to conserve energy

Dependent Variable	Treatment (I)	Treatment (J)	Mean Difference (I-J)	Std. error	p-level
Intentions to conserve energy	Loss frame	Gain frame	-.12	.13	.64
		Control	.69*	.14	<.0001
	Gain frame	Loss frame	.12	.13	.64
		Control	.82*	.13	<.0001
	Control	Loss frame	-.69*	.14	<.0001
		Gain frame	-.82*	.13	<.0001

* Mean Difference is significant at 0.05 level

The second research question deals with the nature of the link between the type of appeal and the LOHAS rating of an individual. To investigate this, only cases that were exposed to environmental messages were selected. The two types of environmental messages that subjects saw dealt with the loss and gain of recycling so a two-way ANOVA with individuals' intentions to recycle more was conducted. Table 5 shows the 2x2 design with the number of subjects in each cell with row and column totals.

Table 5. Number of subjects in each cell of our 2x2 design

	Loss Framing	Gain Framing	Total
LOHAS—high	63	57	120
LOHAS—low	43	59	102
Total	106	116	222

Only the main effect of LOHAS is significant, suggesting that individuals who rated high on the LOHAS scale expressed their intentions of recycling more than those who rated low on LOHAS. To further examine this difference a post-hoc pairwise comparison was made (see Table 6).

Table 6. Two-way ANOVA Dependent variable: Intentions to recycle more

Source	Type III Sum of Square	df	Mean Squares	F-ratio	p-level
Corrected Model	6.323	3	2.108	1.89	.132
Intercept	1793.57	1	1793.573	1608.506	<.0001
LOHAS	5.235	1	5.235	4.695	.031
Ad	1.62	1	1.62	1.453	.229
LOHAS X Ad	.154	1	.154	.138	.711
Error	243.08	218	1.115		
Total	2106.00	222			
Corrected Total	249.40	221			

Table 7a. Mean differences between groups; DV: intentions to recycle

Dependent Variable	Treatment (I)	Treatment (J)	Mean Difference (I-J)	Std. error	p-level
Intentions to recycle more	LOHAS-low	LOHAS-high	-.31*	.143	.031
	LOHAS-high	LOHAS-low	.31*	.143	.031

* Mean Difference is significant at 0.05 level

Table 7b. Means of "intentions to recycle"

Treatment	Mean	Std. Error	95% Confidence Interval Lower Bound	Upper Bound
LOHAS-low	2.718	.106	2.509	2.926
LOHAS-high	3.028	.097	2.838	3.218

From Tables 7a and b, it can be seen that individuals who rated high on the LOHAS scale reported greater intentions to recycle after seeing the environmental messages than the ones low on the LOHAS scale.

Discussion

Interest in the environment is at an all-time high. However, this interest may not always be coupled with behavioral changes that can help the environment. In the coming years, there is likely to be an increased activity in the area of environmental communication research. Issues like message framing, values, and lifestyles of individuals and their role in changing individuals' attitudes and behavior will be increasingly investigated. The current study, though exploratory in nature, provides some evidence for the role of message framing in persuading individuals to act pro-environmentally.

The two main findings of this study are (1) individuals exposed to environmental messages report a greater intention to recyle and conserve energy than those exposed to a non-environmental ad and (2) individuals high on the LOHAS scale are more persuaded to recyle and conserve energy than those rating low on LOHAS. However, these results have shown no interaction between the type of frame and the type of individual. These findings present numerous implications for environmentalists.

First, the fact that environmental messages are influencing individuals to at least report a greater intention to recycle and/or conserve energy is very good news for environmentalists. It shows that the marketing approach has some impact. As discussed earlier, American companies spend over $260 billion on advertising (Boris, 2011). If the persuasion industry can convince us to buy and consume products, the same techniques can also be used to "unsell" and persuade people to conserve and recycle. Contrary to what Crompton (2008) argues, small incremental changes in individuals' behaviors can have an impact on a larger/grander scale much like how consumerism has invaded our collective consciousness over the past century.

Next, if individuals can be persuaded via environmental ads, what types of appeals works best for what types of individuals? The good news is that our results suggest that irrespective of whether individuals saw a loss-frame ad or a gain-frame ad, their response in both cases was a higher level of intention to recycle and conserve energy. The bad news is that there is no empirical evidence in this study to indicate an interaction between the two variables. Both frames appear to work equally well for both types of LOHAS individuals. The role of values and lifestyles in the persuasion process has been suggested to be stronger in activating behavioral changes that ultimately lead to policy changes (Crompton, 2008). It would be ideal for environmentalists to know, for instance, how to target individuals who are low on the LOHAS scale or individuals whose values are more geared toward individuality and materialism. Knowing what appeal (loss or gain) works best for what type of values and lifestyles (intrinsic—personal growth, emotional intimacy, community involvement—or extrinsic—acquisition of material goods, financial success, image and social recognition) will help move the environmental movement forward. Unfortunately, our research failed to establish that link. One reason for this could be that our manipulation changed only the text in the loss or gain manipulations. This type of manipulation may not be big enough to elicit a significantly different response for both cases. Future research needs to investigate whether a simple image (such as a can) or an image that provides a more explicit frame (such as a consumer modeling the desired behavior) evokes a stronger response. Additionally, this study used a single-exposure full-page print advertisement instead of a 30-second television commercial. There have been studies that show that television commercials are more effective in eliciting certain emotions than print advertisements (Schumann, Hathcote & West, 1991). Future research needs to examine the same issues of message framing in multimedia environments.

Our study provides some evidence for the effectiveness of the "marketing approach" to changing behavioral intentions. Given our findings and the general success of the "marketing approach" in creating a culture of consumerism, it may be unwise to discredit the role and importance of such an approach. If the marketing approach is used appropriately, it can bring about changes in pro-environmental behaviors that ultimately result in garnering support for governmental regulation and policy changes.

References

Amidon, J.M. (2005). America's strategic imperative: A 'Manhattan Project' for energy, *Joint Forces Quarterly 39*, 68–77.

Belk, R. (1985). Materialism: Trait aspects of living in the material world. *The Journal of Consumer Research 12*(3), 265–280.

Benefits-of-recycling.com (2010). Recycling statistics, Retrieved online March 21, 2012 from http://www.benefits-of-recycling.com/recyclingstatistics.html

Boris, C. (2011, October 13). Global ad spend expected to hit $500 billion. *Marketing Pilgrim.* Retrieved from http://www.marketingpilgrim.com/2011/10/global-ad-spend-expected-to-hit-500-billion.html

Crompton, T. (2008). Weathercocks & signposts: The environment movement at a crossroads. WWW-UK's Strategies for Change Project. Retrieved online March 21, 2012, from www.valuesandframes.org/downloads

Davis, J. J. (1995). The effects of message framing on response to environmental communications. *Journalism & Mass Communication Quarterly, 72*(2), 285–299.

De Dreu, C. K. & McCusker, C. (1997). Gain-loss frames and cooperation in two-person social dilemmas: A transformational analysis. *Journal of Personality and Social Psychology, 72*(5), 1093–1106.

De Martino, B., Kumaran, D., Seymour, B., & Dolan, R. J. (2006). Frames, biases, and rational decision-making in the human brain. *Science, 313*(5787), 684.

Duhè, S.F., & Zoch, L. M. (1994). Framing the media's agenda during a crisis. *Public Relations Quarterly, 54*(4), 42–45.

Economist Group (2011, February 8). Why don't Americans believe in global warming? *The Economist.* Retrieved from http://www. economist.com/blogs/democracyinamerica/2011/02/climate_change

Entman, R.M. (1993). Framing: Toward a clarification of a fractured paradigm. *Journal of Communication, 43*, 51–58.

Gallup (2010). Americans' global warming concerns continue to drop. Retrieved online January 22, 2012 from http://www.gallup.com/poll/126560/americans-global-warming-concerns-continue-drop.aspx

Hallahan, K. (1999). Seven models of framing: Implications for public relations. *Journal of Pubic Relations Research, 11*(3), 205–242.

Höfer, R. (2009). *Sustainable solutions for modern economies.* Cambridge, United Kingdom: Royal Society of Chemistry Publishing.

Kaufman, L. (2007). Selling green: What managers and marketers need to know about consumer environmental attitudes. *Environmental Quality Management, 8*(4), 11–20.

Lincoln, C. (2009). Light, dark and bright green environmentalism. *Green Daily.* Retrieved from http://www. greendaily.com/2009/04/23/light-dark-and-bright-green-environmentalism/

LOHAS (2008). Lifestyles of Health and Sustainability. LOHAS *Online*. Retrieved online June 22, 2010 from http://www.lohas.com/about.html.

Michaelson, J. (1998). Geoengineering: A climate change Manhattan project, *Stanford Environmental Law Journal, 17*(1998), 73–102.

Moschis, G. (1987). *Consumer socialization: A life-cycle perspective.* New York: Lexington.

Mowery, D.C., Nelson, R.R., & Martin, B.R. (in press). Technology policy and global warming: Why new policy models are needed (or why putting new wine in old bottles won't work). *Research Policy.*

The Nature Conservancy (2010). New Harris Poll shows Americans want to go green, but don't always know how. Retrieved online June 22, 2011 from http://lifestyletom.com/path/rao10925685177ros/roin59012511386

Schumann, D.W., Hathcote, J., & West, S. (1991). Corporate advertising in America: A look at effectiveness studies. *Journal of Advertising, 20*(3), 35–56.

Smith, C. (2010). Insights into recycling behaviors. Retrieved online from http://www.mswmanagement.com/july-august-2010/insights-recycling-behavior.aspx

Tversky, A. & Kahneman, D. (1981). The framing of decisions and the psychology of choice. *Science, 211*(4481), 453.

Williams, A. (2008). "Recycling by the numbers." The good, bad and ugly of statistics and comparisons. Retrieved April 10, 2009 from http://sustainablog.org/2008/08/22/recycling-by-the-numbers-the-good-bad-and-ugly-of-statistics-and-comparisons/

CHAPTER EIGHT

Pro-Environmental Behaviors Through Social Media: An Analysis of Twitter Communication Strategies

Denise Sevick Bortree

This chapter responds to the call by Tom Crompton in "Weather-cocks and Signposts" for an adoption of a values-based approach for promoting environmental behaviors. The study presented here examines the social media communication of 34 Sierra Club groups and measures the use of the marketing approach, alternative approach and a relationship-based approach. Findings suggest that all three are being used, but patterns emerge at different levels of the organization and for different issues and behaviors. Encouraging environmental behaviors is an important activity in environmental public relations, and these findings suggest that advocacy organizations may be consistently selecting less effective strategies to motivate certain behaviors around important environmental issues. Social media communication has become one of the primary channels of communication for advocacy organizations, so careful selection of tactics and strategies could lead to more effective communication and greater impact.

The plight of the environment has been discussed at length in popular and academic literature (Cox, 2010). The causes and solutions are debated, but most agree that something must be done to curb the damage being done to the environment on a daily basis. In his popular white paper "Weathercocks and Signposts," Tom Crompton (2008), Change Strategist for the UK World Wildlife Foundation, argues that the solution starts with changes among governments, businesses, and individuals. As a leader in one of the largest environmental advocacy groups, he recommends that non-profits take an active role in influencing individual behavior through a value-based approach which taps into identity and intrinsic motivations.

One aspect of behavior change that "Weathercocks and Signposts" (Crompton, 2008) does not consider is the role of the relationship that advocacy groups form with their key publics. The public relations model of relationship management purports that through cultivation strategies an organization can improve a relationship that exists between an organization and its key publics (Hon & Grunig, 1999). The quality of the relationship can lead to behavioral outcomes both toward the organization and toward issues.

Advocacy groups use a variety of channels to communicate with audiences and attempt to motivate change. Groups actively promote their causes through social media sites, such as Twitter, Facebook, and LinkedIn. This strategy allows for the creation of dialogue with key stakeholders including donors, volunteers, employees, government, clients, and communities. This chapter adds to the literature on environmental communication by exploring the use of social media among advocacy organizations to motivate environmental behaviors.

Below is a case study that examines the use of the marketing and alternative approaches to behavior change while identifying relationship management strategies that one advocacy group is using to promote and reinforce environmentally responsible behaviors through multiple Twitter streams. The study looks at the differences between public and private behaviors as well as the types of environmental issues that the organizations promote using the different approaches. Results offer insight into the use of Twitter by local, state-wide, regional, and national groups to motivate environmental action.

Literature Review

Behavior motivation has been discussed at length in a broad range of disciplines including health (Moorman & Matulich, 1993; Dutta

& Feng, 2007; Chang, 2009), communication (Krivonos, 1978; Kirzinger et al., 2012; Mead et al., 2012), and psychology (Lee &
Holden, 1999; Sargeant & Shang, 2011; Vinitzky & Mazursky,
2011). Scholars have studied the communication patterns of nonprofit organizations and have offered a wide range of recommendation to strengthen communication and engage with audiences
(Olsen, Keevers, Paul, & Covington, 2001; Wenham, Stephens, &
Hardy, 2003; Waters & Lord, 2009; Taylor, Kent, & White, 2001;
Reber & Berger, 2005; Reber & Kim, 2006).

The following sections will review two approaches that environmental advocacy groups take to encourage environmental behaviors. The first section briefly summarizes the recommendations
from "Weathercocks and Signposts" (Crompton, 2008) which have
been presented elsewhere in this book. The second section makes a
case for examining relationship management as a viable option for
environmental engagement.

Marketing vs. Alternative Approaches

According to Crompton (2008), the most effective way to establish
long-term systemic behavioral change in the environmental realm
is to create a set of values around environmental issues that can
be used as a behavior motivation. Deemed the alternative (values-
based) approach, it addresses the reality that people will abandon
inconvenient behaviors unless they tap into personal values and/or
intrinsic motivations. Only with a strong commitment to environmental behaviors will true changes be made toward environmental
issues. However, Crompton (2008) claims that today's environmental movement is relying too heavily on the marketing approach to
environmental change. The marketing approach is typified by
green consumption and the "small step" approach to large-scale
problems. Unfortunately, there is little evidence that it leads to
sustained change.

The alternative approach to environmental change begins with
a shift in values, but it also hinges on appeals to goals and motivations of individuals. According to Crompton, goals for environmental behaviors can be intrinsic or extrinsic.

...intrinsic goals include personal growth, emotional intimacy or community involvement—they are goals that are inherently rewarding to pursue. Extrinsic goals include acquisition of material goods, financial success, physical attractiveness, image and social recognition. Unlike the intrinsic goals, their pursuit does not lead directly to the satisfaction of innate psychological needs (such as belonging)—rather, the satisfaction they confer is contingent upon the responses of others. (p. 31)

Appealing to intrinsic motivations is one of the eight recommendations made by Crompton (2008) to further promote the alternative approach, and it is one of the few that apply to individual behavior. By appealing to intrinsic motivations, advocacy organizations are able to create communication that leads "to more energetic and persistent audience engagement with environmental issues" (p. 35).

The question of marketing vs. alternative (values-based) approach is particularly important for public relations professionals, who are often employed by advocacy organizations to help promote the organization's mission (Waters et al., 2009; Waters & Jamal, 2011). Public relations has long held that the traditional marketing approaches may be useful for encouraging purchases but are not as effective for making long term changes in socially responsible behaviors. As Crompton argues, individuals may be willing to make small and simple changes through extrinsic motivation, but when behaviors require a great deal of sacrifice, an individual relies heavily on values to guide decisions. One way that nonprofits can engage with audiences to encourage long-term behavior change in public groups is through building and maintaining high quality relationships.

Relationship Management

While Crompton's alternative approach provides an interesting addition to our understanding of behavioral motivation, it does not necessarily consider the contribution of organization relationships that advocacy groups build with publics. "Weathercocks and Signposts" (Crompton, 2008) gives a nod to the social context of values and identity, but it does not allow for the influence of relationship quality that advocacy groups actively build through their interactions with publics. The public relations literature has demonstrated that the quality of the organization-public relationship is critical to the long-term intentions of publics toward the organization (Ki & Hon, 2007a; 2007b; Jo et al., 2004). But, the literature has also suggested that the relationships that nonprofits build with their publics can lead to a change in values as well (Bortree, 2010). The way that organizations communicate to and interact with publics plays an important part in how the organization and its issues are perceived.

Strong organizational relationships are high in trust (one party's confidence that it can be open and honest with another party), control mutuality (balance of power in the relationship), satisfaction (parties involved have positive feelings about one another), and commitment (parties' beliefs that the relationship is worth maintaining) (Ki & Hon, 2007a). These dimensions of the relationship create a bond between organization and public that can lead to long-term commitment and other behaviors (Bortree & Waters, 2010). The organization can take an active role in building the relationship through behaviors and communication strategies toward the publics called cultivation strategies (Hon & Grunig, 1999). The seven cultivation strategies identified in the literature are positivity (making the relationship enjoyable), assurances (legitimizing the concerns of others), shared tasks (working together on a mutually beneficial task), openness (sharing information openly, disclosing), networking (building connections with other entities), access (granting audience access to management/decision makers), and guidance (offering advice, providing help) (Ki & Hon, 2009a; 2009b). A study of volunteers found that the most important cultivation strategies for volunteers were guidance, assurances, and shared tasks (Bortree, 2010). These strategies lead to higher quality in the relationship between volunteers and the organization as well as a change in behavioral intention toward future volunteerism. The study presented here will explore the degree to which cultivation strategies are used in the Twitter streams of environmental advocacy groups and how organizations use the strategies to motivate behavior. The following sections briefly summarize advocacy groups' use of social media.

Social Media and Environmental Advocacy Groups

Environmental advocacy groups use a variety of media and interpersonal channels to communicate their messages to key stakeholders; however, many have moved to social media for active daily promotion of their issues to interested publics including government, sponsors, volunteers, members, and local communities (Lovejoy et al., 2012). Social media sites such as Twitter, Facebook, and LinkedIn provide an avenue for nonprofit groups to connect with their followers and encourage participation in public and private behaviors along a wide range of issues (Curtis et al., 2010; Waters et al., 2009). What isn't known is whether these outlets are conducive for the strategies discussed in the relevant literature and whether organizations are engaging in them.

With the rising popularity of social media, more needs to be done to understand how environmental behavior can be promoted through these outlets.

Characteristics of Twitter

Following is a summary of the characteristics of Twitter, an important social media outlet used frequently by advocacy groups. Twitter is a micro-blogging service with open access, allowing nonregistered viewers to access the site and view the content. Individuals who choose to have a public account post to the site, understanding that their material may be viewed by anyone. Because posts are limited to 140 characters, posts often include markup language such as: "@" preceding a user name, which directs the post to that person; "#", a hashtag, which is used to aggregate topics; and "RT", retweet, which indicates that a post is being passed along from someone else.

Twitter offers organizations the opportunity to engage in a two-way dialogue with their members, and it can be a channel for listening to what stakeholders want from the organization (Briones, Kuch, Liu, & Jin, 2011; McCorkindale, 2010). However, research suggests that nonprofit organizations are not using Twitter to its full potential, with many still using it as a one-way channel to deliver messages (Lovejoy, Waters, & Saxton, 2012).

Levels of Organizational Communication

An important element of social media communication that is ignored in prior research is the psychological distance or proximity between the organization and the audience (Bortree & Dou, 2012). Organizations today manage many different Twitter accounts, and environmental advocacy groups face the challenge of communicating with audiences of local, regional, and national groups. The theory of psychological distance proposes that events and objects that are perceived as more distant are interpreted as more abstract while psychologically close events and objects are interpreted with more detail (Liberman, Trope, & Stephan, 2007). Likely, when an advocacy organization is communicating to a large national audience about large scale issues, the perceived distance between the organization and the audience is greater than when a local advocacy organization is communicating to members who live in a specific city, region or state. In the latter situation, the issues that the local organization promotes may be more relevant and understood in greater detail. Then, too, it is likely that the followers of social media of a local advocacy organization are interested in local issues and the work of the local group. This could have implications for the types of strategies that organizations use to motivate behavior.

To examine the types of strategies that environmental groups use to promote behaviors at different levels of the organization, the following research questions are posed.

RQ1: To what degree are the marketing and alternative approaches used at different levels of organizational communication?

RQ2: To what degree are cultivation strategies used at different levels of organizational communication?

Environmental Behaviors

Social media channels provide environmental advocacy groups a medium to engage with audiences and motivate their behaviors toward environmental issues. Environmental behaviors are often categorized as either public-sphere behaviors or private-sphere behaviors (Stern, 2000; Lee, 2008), with public-sphere behaviors focusing on political or public actions, such as protesting, writing to a congressman, or attending a public event, and private-sphere behaviors focusing on personal actions such as recycling or home energy conservation. Advocacy groups should encourage both types of behaviors in order to achieve substantial environmental changes. In this study public-sphere behaviors are parsed into two groups—those that are organization focused and those that are not related to the organization. Both are public-sphere behaviors, but one directly benefits the organization (donations, volunteering, organization events) as well as the environment. This will allow the study to identify the degree to which advocacy groups are encouraging behaviors toward their organization and the cause vs. toward the cause only.

To explore how environmental behaviors were promoted in the social media, the following research questions were proposed.

RQ3: To what degree are marketing and alternative approaches used to promote public vs. private-sphere behaviors?

RQ4: To what degree are cultivation strategies used to promote public vs. private-sphere behaviors?

Environmental Issues

It is possible that behavior toward some environmental issues are more easily motivated and sustained with one communication strategy or another. Our planet faces many devastating issues, including pollution, greenhouse gas emission, the destruction of species and habitats, and the depletion of energy resources, among others (Cox, 2010). Educating audiences about these issues and connecting their public and private behaviors to the perpetration of environmental concerns requires consistent sustained communication. What is not known is how environmental advocacy organizations promote these issues in their social media communication.

RQ5: Are the marketing and alternative approaches used to promote different environmental issues?

RQ6: Is there a relationship between cultivation strategies used and environmental issues?

Methods

To explore the research questions proposed in this study, a content analysis was conducted on the Twitter streams of 34 Sierra Club groups which were classified as national, regional, statewide, and local. Six months of tweets were captured, which included 3098 tweets. A random sample of approximately 20% (637) was pulled for analysis. The characteristics coded included number of characters, presence of a link, month and date of post, use of RT, and use of @. The posts were coded for their strategy (marketing or alternative approach), relationship cultivation strategy, level of organizational communication, behavior type (personal-sphere, public-sphere, or organization-focused), and environmental issue.

Coding. Two coders analyzed the sample after first establishing reliability through a sample of 50 tweets. Overall, the intercoder reliability for the variables was 94%, which falls within the acceptable range for reliability.

Strategy (Marketing vs. Alternative Approach). Crompton (2008) suggested that the ideal way to ensure sustained environmental change is through creating and appealing to values people hold in relationship to the environment. He dubbed this the alternative approach, in contrast to the marketing approach, which holds that small steps can lead to large changes. To measure the use of these two strategies (marketing vs. alternative approach) in social media communication, persuasive strategies were first identified and then categorized into one of the two approach types. A thorough review of the sample in this study suggested eight strategies that are used to encourage environmental behaviors. These strategies were divided into two groups, one that reflected a marketing approach and one that reflected the alternative approach.

Four of the eight strategies were grouped into the marketing approach. They included: *compensation* (offering some kind of give away or payment or free attendance to an event as a motivation to get the reader to attend), *enjoyment/fun/interest* (suggesting that the reader will have fun, enjoy oneself, or find the event or behavior interesting), *professional fulfillment* (suggesting that the reader will have some professional success or will take steps that might help in some professional aspect of one's life), *peer pressure/social group pressure* (suggesting that others/peers are taking this action or that certain behaviors are indicative of "others" not accepted with a social group), These were deemed to be part of the marketing approach because they appeal to personal gain rather than focus on making a contribution to the environment through intrinsic motivations.

The other four strategies were grouped into the alternative approach, including: *personal fulfillment* (suggesting that the reader will find personal fulfillment by taking action), and *becoming more knowledgeable/education* (suggesting that one will learn and/or become more knowledgeable by taking an action), *raise awareness of issue/org* (suggesting that the outcome of the behavior will lead to greater awareness of the organization or the issue) and *contribute to cause/org* (suggesting that the outcome of the behavior or action will lead to a contribution in some way to the organization or cause beyond the simple awareness suggested in the prior option). These strategies framed environmental actions as having long-term significant consequences for the environment or relying on intrinsic motivations.

Relationship Cultivation Strategies. To examine whether advocacy groups are using Twitter to build relationships with their audiences, posts were coded for their use of relationship cultivation strategies as defined by Ki & Hon (2009a; 2009b). Strategies included positivity, assurances, shared tasks, openness, networking, access, and guidance. Each post was coded for its dominant cultivation strategy, if more than one were present.

Level of Organizational Communication. The 34 Twitter streams included streams from four different levels of the Sierra Club organization: local (city-level group or area smaller than a state), state (state-level group), regional (group for area larger than one state but not national), and national (national-level group). Each post was coded for the level of the organization that issued it.

Behavior Type. Posts that suggested a behavior were coded for behavior type. Behaviors were grouped into three categories: *personal-sphere behavior* (any personal action toward the environment including recycling, taking a class, or reduction of water use), *public-sphere behavior* (public activities such as coordinating a rally or event, picketing, contacting a congressman, etc.), and *organization-focused behavior* (any public-sphere activity that is facilitated by the organization or any actions that involve the organization such as donating money, volunteering, attending the organization's event, etc.).

Environmental Issue. Posts were reviewed for the primary environmental issue that they promoted or addressed. Issues that were coded including pollution, greenhouse gas reduction, solid waste problems, species/habitat preservation, energy efficiency, energy independence, and the general state of the economy. If more than one environmental issue was raised in a post, an attempt was made to identify the primary issue.

Results

A total of 637 tweets were analyzed in the study. Their average length was 112 characters (SD = 28); approximately 85% included links; 12% were retweets; and 18% were responses to other posts. Just over half of the tweets were posted by state level groups (55%); 30% of tweets came from a national Twitter stream; 9% were from local groups; and 6% were from regional groups.

The most popular environmental issues were pollution (25% of posts), general state of the environment (19%), species/habitat preservation (19%), and energy independence (10%). Approximately 12% of the tweets did not address a specific environmental issue. The least popular environmental issues were greenhouse gas reduction (2%), solid waste problems (4%), and water conservation (4%).

Approximately, 46% of the posts promoted a behavior; 42% delivered information; and 12% attempted to influence an attitude of the reader. Because this study examines behavior motivation, only the posts that promoted a behavior (n = 294) were used to answer the research questions.

Research Questions

Research question one asked about the use of marketing vs. alternative approaches at the different levels of the Sierra Club. More posts used the values-based approach (n = 178) than the alternative approach (n = 116) in this study. A chi-square test was run, and the results suggested that the distribution was not significantly different than expected (x^2 (3, n = 294) = 3.81, p = .28). Therefore, it is assumed that the approach used by the different levels of the organization was about what would be expected if the two variables had no significant relationship (Table 1).

Table 1: Chi-square of marketing vs. alternative approaches at different organizational levels

Org Level		Marketing	Alternative	Total
National	Observed	19	43	62
	Expected	25	37	62
Regional	Observed	9	18	27
	Expected	11	16	27
State	Observed	71	98	169
	Expected	67	102	169
Local	Observed	17	19	36
	Expected	14	22	36
	Total	116	178	294

(x^2 (3, n = 294) = 3.81, p = .28)

The second research question asked about the use of relationship management cultivation strategies at the different levels of the organization. Only three relationship strategies were observed frequently enough to have at least five predicted observations at each level of the organization. The strategies were positivity, shared tasks, and guidance.

Two of these strategies were identified in the literature as important for volunteers (Bortree, 2010), so their presence in social media suggests that they are commonly used for other nonprofit stakeholders as well. A chi-square analysis suggested that the observed frequencies of these strategies were significantly different from expected (x^2 (6, n = 260) = 17.43, p < .01). At the national level, the organization tended to use positivity and guidance slightly more than expected and shared tasks less than expected. At the state level, the strategy of shared tasks was used more often than expected, and guidance was used less than expected. At the local level, the organization tended to use guidance more than expected and positivity and shared task not as often as expected. Strategies were used as expected at the regional level groups (Table 2).

Table 2: Chi-square cultivation strategies used at different levels of the organization

Org Level		Positivity	Shared Tasks	Guidance	Total
National	Observed	25	13	28	66
	Expected	20	21	26	66
Regional	Observed	7	7	9	23
	Expected	7	7	9	23
State	Observed	41	58	48	147
	Expected	44	46	57	147
Local	Observed	5	3	16	24
	Expected	7	8	9	24
	Total	78	81	101	260

(x^2 (6, n = 260) = 17.43, p < .01)

The third research question asked about the use of marketing vs. alternative approach as a strategy for promoting public vs. private-sphere behaviors. A chi-square crosstab suggested that the results are significantly different from expected. Interestingly, the organization uses the marketing approach more than expected when motivating behavior toward the organization, but it uses the alternative (values) approach more often when encouraging other public-sphere behaviors and private-sphere behaviors. (χ^2 (2, n = 290) = 45.68, p < .001) (Table 3).

Table 3: Chi-square of private vs. public-sphere behaviors promoted via marketing vs. alternative approaches

Behavior Type		Marketing	Alternative	Total
Private sphere	Observed	13	47	60
	Expected	24	36	60
Public sphere	Observed	10	58	68
	Expected	27	41	68
Org-related (public sphere)	Observed	92	70	162
	Expected	64	98	162
	Total	115	175	290

(χ^2 (2, n = 290) = 45.68, p < .001)

Research question four looked at the use of cultivation strategies for promoting public vs. private-sphere behaviors. Using the same three strategies, a chi-square crosstab found that expected and observed frequencies were significantly different (χ^2 (4, n = 207) = 31.60, p < .001). Positivity was used less than expected with public-sphere behaviors and more than expected with organization-related behaviors. Shared tasks was used less than expected with private-sphere behaviors and more than expected with public-sphere behaviors. Guidance was used more than expected related to private-sphere behaviors and less than expected for the other two behavior types (Table 4).

Table 4: Chi-square cultivation strategies used at different levels of the organization

Behavior Type		Positivity	Shared Tasks	Guidance	Total
Private-sphere	Observed	10	4	32	46
	Expected	9	17	20	46
Public-sphere	Observed	3	31	19	53
	Expected	10	20	23	53
Org-related (public sphere)	Observed	25	43	40	108
	Expected	20	41	48	108
	Total	38	78	91	207

$(\chi^2 (4, n = 207) = 31.60, p < .001)$

The fifth research question looked at the use of marketing and alternative approaches to promote different environmental issues. Only four issues were observed enough to be used in a chi-square crosstab calculation. For each, at least 5 observations were expected in each cell. The four issues were pollution, general state of the environment, species/habitat, and energy independence. The results suggested that the observed frequencies were significantly different from expected $(\chi^2 (3, n = 218) = 31.67, p < .001)$. The organization used the marketing approach more often than expected when focusing on species/habitat and energy independence, and it used the alternative approach more than expected when discussing pollution. For general state of the environment, the use of both approaches matched the expectation (Table 5).

The final research question asked about the use of cultivation strategies for promoting environmental issues. Using the top four environmental issues, and the three cultivation strategies mentioned above, a chi-square analysis was run to test whether the observed and expected frequencies were significantly different. Results suggest that the differences are significant $(\chi^2 (6, n = 162) = 27.6, p < .001)$. Pollution was discussed using all strategies approximately as expected. Species/habitat preservation was discussed significantly less using shared tasks than expected and somewhat more than expected using positivity and guidance (Table 6). Energy independence is promoted more than expected using shared tasks and less than expected with guidance. The general state of the environment is discussed more than expected with guidance and less than expected using shared tasks.

Table 5: Chi-square of marketing vs. alternative approaches to environmental issues

Environmental Issue		Marketing	Alternative	Total
Pollution	Observed	8	50	58
	Expected	23	35	58
Species/habitat	Observed	28	30	58
	Expected	23	35	58
Energy independence	Observed	26	12	38
	Expected	15	23	38
General state of environment	Observed	23	41	64
	Expected	25	39	64
	Total	85	133	218

$(\chi^2 (3, n = 218) = 31.67, p < .001)$

Table 6: Chi-square of cultivation strategies used to promote environmental issues

Environmental Issues		Positivity	Shared Tasks	Guidance	Total
Pollution	Observed	9	19	15	43
	Expected	9	16	18	43
Species/habitat	Observed	13	10	23	46
	Expected	9	18	19	46
Energy independence	Observed	3	26	7	36
	Expected	7	14	15	36
General state of environment	Observed	7	9	21	37
	Expected	7	15	15	37
	Total	32	64	66	162

$(\chi^2 (6, n = 162) = 27.6, p < .001)$.

Discussion

This study found that groups of one environmental advocacy organization commonly use marketing and alternative approaches to promote environmental behaviors on its Twitter stream, with the alternative approach being used more often than the marketing approach. This is encouraging and suggests that environmental advocacy groups may see the value of the alternative approach for promoting environmental behaviors. At the same time, the groups appear to be using relationship management cultivation strategies to engage with readers and encourage the adoption of environmentally-friendly behaviors, most often using guidance, shared tasks, and positivity. Prior research suggests that guidance and shared tasks play a critical role in the volunteer-nonprofit organization relationship (Bortree, 2010), and finding them in use in advocacy social media communication is a sign that groups are actively engaging them through multiple sources. However, while the groups are attempting to create and build relationships, the analysis suggests that they are not aggressively creating a two-way dialogue, considering that only 12% of posts were retweets and 18% were responses (or used @ to direct a message to another user). The behaviors that the groups promote are twice as likely to be organization focused than either public-sphere or private-sphere, and the behaviors are frequently related to addressing pollution, species/habitat preservation, energy independence and the general state of the environment.

The study looked at the differences in strategy of the Sierra Club groups at different levels of the organization. While the uses of marketing and the alternative approaches were not significantly different from expected, the cultivation strategies did appear to vary significantly from level to level. At the national level, the groups used shared tasks less than expected which suggests that the national groups are not inviting readers to work with them on issues as often as expected. This may be due to the distance perceived between national groups and their followers. Individuals may follow a national advocacy group to learn about an issue but may be less motivated to work with the organization in some capacity. By contrast, at the state level, more shared tasks strategy was used than expected. State-level groups are more actively inviting their members to engage in mutual tasks with them. At the state level there is less guidance than expected, but more than expected at the local level, indicating that local-level groups are offering specific information about how to get involved or how to adopt a behavior. The use of these three strategies may suggest dimensions of the cultivation strategies not explored before. In this context it appears that distance plays a role in the strategy chosen to promote behaviors. Guidance may be appropriate or more effective with publics that have close relationships with an organization or are more likely to have an interpersonal relationship with leaders in the organization. Guidance is used to help individuals understand a process, be more effective, or learn a new behavior, and so the strategy may be more effective when exchanged between people who work together on a local level. Prior research found that organizations that offer guidance to volunteers have stronger relationships with them (Bortree, 2010). Moving one step away from the local groups, the state-level groups were more likely to encourage readers to adopt an environmental behavior to mutually benefit the organization and the individual. States have their own unique environmental concerns, and it appears that these groups were tapping into the need for residents to join the organization in addressing these issues. At the national level, positivity was used more than expected. This strategy is a way for organizations to appear friendly toward publics, and considering the distance between readers and a national-level group, the use of positivity may be a way to try to draw people closer to the group. It is likely that followers of a national stream are less invested in the organization than those who seek out the local or state-level groups. These findings raise questions about the effectiveness of strategies in different contexts, including the distance between publics and organizations.

The study explored the types of behaviors that were promoted, categorizing them as private-sphere, public-sphere, or organization-related. Results suggested that private-sphere and public-sphere behaviors were more likely than expected to be promoted using the alternative approach, and the organization-related behaviors were promoted more often than expected using the marketing approach. These findings raise an interesting question. Crompton (2008) suggests that the marketing approach's simple steps do not lead to big changes for the environment, but does this approach work for an organization that wishes to attract members or encourage greater involvement with one of its groups? Presumably the more support an environmental organization receives, the more good it can do for the environment. Could the marketing approach be a valid way to introduce individuals to environmental issues by first encouraging an association with an organization, and later using the alternative approach to promote subsequent behavior (public-sphere behaviors)?

The cultivation strategies used to promote private-sphere, public-sphere, and organization-related behaviors varied. Guidance was used more than expected to encourage private-sphere behaviors, suggesting that the groups were building a relationship with the reader by offering advice on how to adopt personal environmental behaviors. Shared tasks strategy was used more than expected with public-sphere behaviors. It appears that the groups were asking readers to join them in their public actions toward the environment. Interestingly, the groups used positivity to encourage behaviors toward the organization. When the groups asked readers to act toward the organization, they attempted to portray themselves as friendly. This, too, raises an interesting question about the relationship between behaviors and cultivation strategies. Are individuals more likely to take public actions that benefit an issue when they feel they are engaging in a mutual task with an organization? And, what role does the appeal of the organization play in the adoption of behaviors?

The four environmental issues that were most frequently raised by the Sierra Club groups were pollution, species/habitat preservation, energy independence and the general state of the environment. Pollution was promoted more than expected using the alternative approach, and the issues species/habitat preservation and energy independence were discussed more often than expected using the marketing approach. This suggests that the groups were appealing to higher-order values when encouraging readers to take actions toward pollution, but actions toward species/habitat preservation and energy independence were encouraged through appeals to personal gain. Species/habitat preservation is the hallmark issue of the Sierra Club; to encourage effective actions toward this issue is a priority for the organization. Energy independence encompasses not only the political issue of independence from foreign resources but also the sources of energy that are mined and used domestically. As Crompton (2008) suggests, the marketing approach likely will lead to temporary changes, and considering the critical nature of species/habitat preservation and energy independence, the Sierra Club groups should consider greater adoption of the alternative strategy to encourage significant changes in values and attitudes of members toward these issues.

The posts encouraging behaviors toward environmental issues used all three relationship cultivation strategies—positivity, shared tasks, and guidance. Actions toward energy independence were presented as an opportunity for shared tasks with the organization while shared tasks was used less than expected with species/habitat preservation. Behaviors toward the general state of the environment were encouraged more often than expected using guidance. At first it might not be clear how behaviors could be tied to the general state of the environment and how guidance might be offered to this end; a closer examination of the Tweets suggested that many of the posts offered links to websites with information on vague topics such as how to "be green" or "celebrate Earth Day." These findings suggest that the groups are engaging in relationship building through shared task activities to address energy independence and positivity toward actions for species/habitat preservation.

Taken together, the findings in this study suggest that Sierra Club groups were using the marketing approach toward organization-related behavior, species/habitat preservation, and energy independence while using the alternative approach private-sphere, public-sphere behavior and the issue of pollution. While Crompton (2008) rejects the use of the marketing approach as a strategy for encouraging environmental behaviors, there may be some validity for its use toward organization-behaviors, or even behaviors toward certain issues, such as energy independence.

Conclusion

This study raises interesting questions about the way environmental groups use social media to encourage environmental behaviors. Future research should test the impact of these messages and strategies to educate and motivate behavior among audiences. The literature would also benefit from an examination of the patterns of Twitter activity among active and aware audiences. It is possible that certain strategies are more effective for active audiences or for aware audiences.

Limitation. The study employed a content analysis of Twitter posts for 34 Sierra Club groups. While the large number of groups suggests variety in posts and topics, the fact that all groups were a part of the Sierra Club limits the generalizability of findings. In addition, the marketing approach vs. alternative approach lacks a clear explication in the literature, and while the authors assumed accuracy in measurement, the literature would benefit from a clearer definition of the approaches.

References

Bortree, D.S. (2010). Exploring adolescent-organization relationships: A study of effective relationship strategies with adolescent volunteers. *Journal of Public Relations Research. 21*(1), 1–25.

Bortree, D.S. & Dou, X. (2012). The role of proximity in advocacy communication: A study of Twitter posts of the Sierra Club Groups, in S. Duhe (Ed.) *New Media & Public Relations* (2nd edition). New York: Peter Lang.

Bortree, D. S. & Waters, R.D. (2010). The impact of involvement in the organisation-public relationship: Measuring the mediating role of involvement between organisation behavior and perceived relationship quality. *PRism 7*(2): http://praxis.massey.ac.nz/prism_on-line_journ.html.

Briones, R. L., Kuch, B., Liu, B. F., & Jin, Y. (2011). Keeping up with the digital age: How the American Red Cross uses social media to build relationships. *Public Relations Review, 37*(1), 37–43.

Chang, C. (2009). Psychological motives versus health concerns: Predicting smoking attitudes and promoting antismoking attitudes. *Health Communication, 24*(1), 1–11.

Cox, R. (2010). *Environmental communication and the public sphere* (2nd ed). Thousand Oaks, CA: Sage Publications.

Crompton, T. (2008). Weathercocks and signposts: The environmental movement at a crossroads. Retrieved from http://www.valuesandframes.org/downloads.

Curtis, L., Edwards, C., Fraser, K.L., Gudelsky, S., Holmquist, J., Thornton, K., & Sweetser, K. D. (2010). Adoption of social media for public relations by nonprofit organizations. *Public Relations Review, 36*(1), 90–92.

Dutta, M. J., & Feng, H. (2007). Health orientation and disease state as predictors of online health support group use. *Health Communication, 22*(2), 181–189.

Hon, L. C., & Grunig, J. E. (1999). *Measuring relationships in public relations.* Institute for Public Relations, Gainesville, FL.

Jo, S., Hon, L. C., & Brunner, B. R. (2004). Organisation-public relationships: Measurement validation in a university setting. *Journal of Communication Management, 9*(1), 14–27.

Ki, E.-J., & Hon, L. C. (2007a). Reliability and validity of organization-public relationship measurement and linkages among relationship indicators in a membership organization. *Journalism & Mass Communication Quarterly, 84*(3), 419–438.

Ki, E.-J., & Hon, L. C. (2007b). Testing the linkages among the organization-public relationship and attitude and behavioral intentions. *Journal of Public Relations Research, 19*(1), 1–23.

Ki, E.-J., & Hon, L. C. (2009a). A measure of relationship cultivation strategies. *Journal of Public Relations Research, 21*(1), 1–24.

Ki, E.-J. & Hon, L.C. (2009b). Causal linkages between relationship cultivation strategies and relationship quality outcomes. *International Journal of Strategic Communication, 3*(4), 242–263.

Kirzinger, A. E., Weber, C., & Johnson, M. (2012). Genetic and environmental influences on media use and communication behaviors. *Human Communication Research*, 38(2), 144–171.

Krivonos, P. D. (1978). The relationship of intrinsic-extrinsic motivation and communication climate in organizations. *Journal of Business Communication, 15*(4), 53-65.

Lee, K. (2008). Making environmental communications meaningful to female adolescents. *Science Communication, 30*(2), 147–176.

Liberman, N., Trope, Y., & Stephan, E. (2007). Psychological distance. In A. W. Kruglanski, & E. T. Higgins (Eds.), *Social psychology: Handbook of basic principles* (Vol. 2, pp. 353–383) New York, NY: Guilford Press.

Lee, J. A., & Holden, S. J. S. (1999). Understanding the determinants of environmentally conscious behavior. *Psychology & Marketing, 16*(5), 373–392.

Lovejoy, K., Waters, R.D., Saxton, G. (2012). Engaging stakeholders through Twitter: How nonprofit organizations are getting more out of 140 characters or less. *Public Relations Review, 38*(2), 313–318.

McCorkindale, T. (2010, August). Twitter me this, Twitter me that: A quantitative content analysis of the 40 Best Twitter Brands. Paper presented at the annual convention of the Association for Education in Journalism and Mass Communication, Denver, CO.

Mead, E., Roser-Renouf, C., Rimal, R. N., Flora, J. A., Maibach, E. W., & Leiserowitz, A. (2012). Information seeking about global climate change among adolescents: The role of risk perceptions, efficacy beliefs, and parental influences. *Atlantic Journal of Communication, 20*(1), 31–52.

Moorman, C., & Matulich, E. (1993). A model of consumers' preventive health behaviors: The role of health motivation and health ability. *Journal of Consumer Research, 20*(2), 208–228.

Olsen, M., Keevers, M.L., Paul, J., & Covington, S. (2001). E-relationship development strategy for the nonprofit fundraising professional. *International Journal of Nonprofit and Voluntary Sector Marketing, 6*(4), 364–373.

Reber, B.H., & Berger, B.K. (2005). Framing analysis of activist rhetoric: How the Sierra Club succeeds or fails at creating salient messages. *Public Relations Review, 31*(2), 185–195.

Reber, B.H., & Kim, J.K. (2006). How activist groups use websites in media relations: Evaluating online press rooms. *Journal of Public Relations Research, 18*(4), 313–333.

Sargeant, A., & Shang, J. (2011). Bequest giving: Revisiting donor motivation with dimensional qualitative research. *Psychology & Marketing, 28*(10). 980–997.

Stern, P.C. (2000). Toward a coherent theory of environmentally significant behavior. *Journal of Social Issues, 56*(3), 407–424.

Taylor, M, Kent, M.L., White, W.J. (2001). How activist organizations are using the Internet to build relationships. *Public Relations Review, 27*(3), 263–284.

Vinitzky, G., & Mazursky, D. (2011). The effects of cognitive thinking style and ambient scent on online consumer approach behavior, experience approach behavior, and search motivation. *Psychology & Marketing, 28*(5), 496–519.

Waters, R.D., Burnett, E., Lamm, A., & Lucas, J. (2009). Engaging stakeholders through social networking: How nonprofit organizations are using Facebook. *Public Relations Review, 35*(2), 102–106.

Waters, R.D., & Jamal, J.Y. (2011). Tweet, tweet, tweet: A content analysis of nonprofit organizations' Twitter updates. *Public Relations Review, 37*(3), 321–324.

Waters, R.D., & Lord, M. (2009). Examining how advocacy groups build relationships on the Internet. *International Journal of Nonprofit and Voluntary Sector Marketing, 14*(3), 231–241.

Wenham, K., Stephens, D., & Hardy, R. (2003). The marketing effectiveness of UK environmental charity websites compared to best practices. *International Journal of Nonprofit and Voluntary Sector Marketing, 8*(3), 213–223.

CHAPTER NINE

Evaluating the Ethicality of Green Advertising: Toward an Extended Analytical Framework

Lee Ahern

Is it ethical for advertisers to use environmental appeals to promote consumption? Is it ethical for environmental advocates to use non-conscious, indirect advertising techniques to promote pro-environmental behaviors? These are two basic questions inherent in the debate over the marketing approach to environmental communication as articulated by Tom Crompton in "Weathercocks and Signposts." What is needed to explore these issues is not an ethics of environmental advertising, but an ethics of advertising better suited to addressing environmental messaging. Researchers and practitioners alike understand that advertising has the potential to deny or suppress individual autonomy in indirect and non-conscious ways. Because these techniques do not involve direct information or logical argument, traditional ethical frameworks don't apply. By more precisely describing the spectrum of indirect advertising techniques, and situating the ethical debate in the context of the public sphere, this chapter puts forth an extended framework for the ethical evaluation of indirect, non-conscious approaches to advertising that informs current debates in the area of environmental communication.

A BP ad from a few years ago (pre-spill) featured a carload of animated children happily driving through a green landscape forested with gently spinning turbines. When the gas gauge gets a bit low they shun a number of dingy, gray gas stations before turning in under the green-sunburst BP logo where their car is quickly filled by a smiling green gas pump. After they drive off into the

BP-logo sunset, the company's Beyond Petroleum tagline appears on the screen.

A more recent Nissan Leaf ad features a polar bear departing the Arctic for a hazardous trek to suburbia. After a number of trials and travails, the bear finds the driveway he has been looking for. A man leaves his house for work and is about to get into his Leaf when the bear comes around the back of the car, rises up on its rear legs—and gives him a hug.

What are we to make of these types of messages? Are they both accurate? Truthful? Fair? The questions don't seem to apply, and debate continues about the effects and ethicality of these types of indirect messages. Many critics would classify both ads as reprehensible greenwashing, while many others would consider both ads harmless image appeals. While it is tempting to lump both ads together as "good" or "bad," there do seem to be some fundamental differences that separate them. Therefore, a flexible and systematic ethical framework for evaluating these types of messages would be useful.

In "Weathercocks and Signposts," Tom Crompton questions the long-term practicality, and unintended consequences, of using marketing appeals to promote pro-environmental behavior. This analysis speaks not only to the dangers of using environmentalism in the pursuit of marketing, but the problematic nature of using marketing in the pursuit of environmentalism. By more precisely describing the spectrum of indirect advertising techniques, and situating the ethical debate in the context of the public sphere, this chapter seeks to do just that.

Advertising Ethics

Ethical evaluations are by definition subjective, and, ultimately, every individual needs to make up their own mind. But because these types of ads are developed by social organizations and broadcast to mass publics, there is a collective ethical responsibility to develop frameworks to guide behavior (Spence & Van Heekeren, 2005). These frameworks take the form of torts and legislation, government regulations, industry codes of ethics, media clearances, and organizational policies. Essentially, these are sets of rules for the classification of certain types of advertising approaches and/or techniques as ethically problematic. Their purpose is to inform the evaluation of individual ads, not to critique the advertising industry or market economics as a whole.

In the United States, these government, industry and self-regulatory policies comprise thousands of pages. The Advertising and Marketing section of the Federal Trade Commission's Bureau of Consumer Protection (http://business.ftc.gov/advertising-and-marketing) lists links to 430 legal resources (cases studies, compliance rulings, laws, etc.) and 106 other documents, many of which are thousands of words long, under the general headings of Marketing Basics, Children, Online, Health and Environment. There are additional guidelines for specific issue areas; the FTC's official Guides for the Use of Environmental Marketing Claims total 7,500 words.

One can also consult the 13 pages of compliance procedures published by the National Advertising Review Council, the main self-regulatory body for the advertising industry. These procedures outline how the NARC handles complaints related to the Better Business Bureau's 13-page Code of Advertising. For more information on specific issues, one can search the 6,000+ cases and findings of the NARC's main operating units—the National Advertising Division, the National Advertising Review Board, the Children's Advertising Review Unit, and the Electronic Retailing Self-Regulatory Program (http://www.narcpartners.org). Then there are the ethics codes from the main industry associations—the American Advertising Agency Association, the Association for National Advertisers, the American Marketing Association, and the Public Relations Society of America—and internal ethics policies for every major marketer and media organization in the world.

These codes and guides are voluminous and are grounded in hundreds and even thousands of years of ethical reasoning and moral philosophy. But based on recent advances in the understanding of human ethical intuition, and the latest social science on how advertising and persuasive messages work, they can be seen as fighting old ethical battles and as focusing the least on those aspects of advertising that matter the most.

Defining Advertising

Advertising refers to any 1) paid, 2) non-personal message from 3) an identified source that is 4) designed to persuade (Belch & Belch, 2009). Alternative definitions abound, and advertising is often seen as synonymous with propaganda, which certainly has more negative connotations (the relationship of advertising to propaganda is explored more closely in the following sections). Proponents of advertising prefer to equate it with "information," which is hard to find objectionable. A more accurate definition is more likely to be found somewhere in the middle, and the four simple elements outlined above make for a useful set of boundaries, if not a universally accepted definition.

Each of the four elements that define advertising has well-known ethical implications (Spence & Van Heekeren, 2005). The paid nature of advertising messages, for example, invites more rigorous ethical scrutiny. The profit motive is critical to modern commerce but can also represent a corrupting influence. It may be difficult for individuals within organizations with strong profit motives to resist institutional pressures to lower their ethical standards. The non-personal aspect of advertising means there is no direct and immediate opportunity for feedback, which is not ideal for open dialog. The ethical implications of obscuring the source of a message are well known, and the persuasive intent of advertising is problematic for certain susceptible audiences (children and the mentally impaired) who do not have a developed sense of "persuasion knowledge." These ethically problematic and inherent aspects of advertising are for the most part effectively addressed by existing regulations and codes of ethics.

The ethical evaluation of informational advertising is relatively straightforward. Arguments are presented and evidence is offered as to why a product/brand fulfills a specific need, and consumers (and regulators) have adequate material with which to evaluate the truthfulness and accuracy of the message. The complex set of regulations, guidelines and codes of ethics that has been built up over decades is well suited for the ethical evaluation of such messages. But a great deal of advertising, especially in the case of national consumer brand advertising, takes a different, non-informational approach.

Alternative Routes to Persuasion

Modern commercial messages use imagery and loose associations to suggest that products or brands will fulfill not just some concrete need but some higher-level desire such as happiness, sexual desirability, or social power. No specific arguments or evidence are offered; the relationships are simply presented as given. Within the advertising ethics literature, such indirect approaches have been variously referred to as indirect information transfer (Arrington, 1982), persuasive advertising (Crisp, 1987; Santilli, 1983; Waide, 1987), associative advertising (Waide, 1987), or self-identity/image appeals (Bishop, 2000). In the professional lexicon, these types of ads are also referred to as transformational, emotional or values appeals (Belch & Belch, 2009).

Indirect advertising can take a variety of forms. It must also be noted that many, if not most, ads are a combination of direct and indirect appeals, and it is difficult to clearly distinguish between the two (Santilli, 1983). A hierarchy of advertising appeal types can be employed to classify messages along a continuum from concrete to abstract focus. Often called the product benefits ladder (Belch & Belch, 2009), concrete product attributes are at the bottom, followed by product benefits or features, consumer benefits and finally values. Ads that focus on the lower and middle parts of the spectrum—product or brand attributes, features and benefits—would fall into the informational classification. Messages that engage more abstract consumer benefits and ultimately values often rely on indirect image advertising. While many ads employ both informational and indirect image appeals, one aspect or approach is usually dominant, and both aspects should be ethically evaluated.

It is significant that indirect image advertising is conceived as engaging in appeals to *values*. Most of the associations created by marketers in image advertising seek to equate a product or brand with some terminal value (a common, socially shared desirable end-state of existence) or instrumental value (preferable modes of behavior or means of achieving terminal values) (Rokeach, 1968). The advertisements referenced at the beginning of this chapter, for example, evoke the terminal value of a sustainable environment ("a world of beauty" in the lexicon of the original Rokeach Values Survey). Careful examination of an indirect image ad will yield a dominant value that is being evoked, and doing so is an important step in its ethical evaluation.

Apart from the use of indirect values/image advertising, there is a spectrum of indirect advertising techniques that can be seen as moving advertising into a related, and more ethically problem-atic, realm—that of propaganda. Following an outline of the dis-tinctions between advertising and propaganda, and a review of how existing ethics codes approach this difficult area, examples are provided to more specifically describe the spectrum of indirect, non-conscious advertising techniques. These techniques are then reviewed in the context of literature on ethics and the unconscious in general, and the ethics of non-conscious advertising in particu-lar.

Advertising and Propaganda

Propaganda can be defined as "a systematic, motivated attempt to influence the thinking and behavior of others through means that impede or circumvent a propagandee's ability to appreciate the na-ture of this influence" (Marlin, 2003). Given this definition, there is certainly an overlap between propaganda and advertising. All advertising satisfies the first part of the definition: "a systematic, motivated attempt to influence thinking and behavior." Not all ad-vertising, however, operates through "means that impede" persua-sion awareness. The traditional informational approaches to advertising persuasion basically offer syllogistic arguments about product or brand superiority. For these types of messages, advertisers don't want to impede the audience's ability to evaluate the persuasion. Quite the opposite. In order for purely informational messages to have persuasive impact, the audience needs to selectively perceive, comprehend and evaluate the arguments and, hopefully, come down on the side of the marketer. In the context of dual-process models of persuasion, they need to centrally or systematically process the information (Chaiken, Libermann, & Eagly, 1989; Fishbein & Ajzen, 1975).

For a variety of reasons, however, audiences do not generally commit the cognitive effort to systematically process advertising messages (Sutherland & Sylvester, 2000). Consumers see too many ads for too many products (most of which they have no interest in) to pay attention to them all. For many product categories brands are seen as interchangeable; most consumers are not going to spend much time thinking hard about competing brands of all-purpose flour. Even for high-involvement products like flat-panel TVs, most audience members are not in the market at any particular time and are not going to give the details of the sales pitch too much thought. Marketers understand this. To overcome this kind of limited attention and low cognitive involvement, they often seek to impart brand information into the minds of consumers with techniques that can be considered efforts to impede persuasion awareness. Many of these techniques are relatively benign, while others might be considered more problematic.

An advertisement that satisfies the definition of propaganda by virtue of the techniques used to deliver it is not by definition unethical, but it can certainly be seen as entering an ethically problematic area. Ironically, marketing regulations and codes of advertising ethics focus predominantly on the traditional, syllogistic-informational ads that audiences have a natural ability to defend and offer little guidance on how to evaluate messages that employ more complex and subtle psychological techniques.

The Better Business Bureau's Code of Advertising is representative of most attempts to detail fair and accurate advertising practices. It comprises long lists of specific types of informational claims and arguments that may be inappropriate. Section headings include "Comparative Price, Value and Savings Claims," and "Bait Advertising and Selling," etc. Nowhere does this framework address the use of indirect association, implications, imagery, narrative or emotion. The only section that looks at subjective aspects of advertising is "Superlative Claims-Puffery." Puffery is the legal term for general claims that people clearly recognize as hyperbole ("best," "ultimate," "favorite," etc.) and which is not prohibited or seen as ethically problematic. Here the Code merely states that subjective claims "may sometimes be considered puffery and not subject to tests of their truth and accuracy," and "subjective superlatives which tend to mislead should be avoided." This language largely limits the application of the Code to advertising with verifiable claims. The only non-verifiable aspect of advertising subject to evaluation would be potentially misleading superlative claims.

The BBB's Code of Advertising and others like it are fine frameworks for the evaluation of direct and informational advertising with objectively verifiable claims. Another level of analysis is required, however, to address the ethical implications of more indirect advertising techniques.

The Spectrum of Non-Conscious Routes to Persuasion in Advertising

Advertising giant David Ogilvy often made the case that the best ads are those that sell *without drawing attention to themselves* (Ogilvy, 1985). He understood that the best way to sell something is to somehow distract people from the fact that you are trying to sell them something. Although this approach is counterintuitive for personal selling, it makes enormous sense for building and positioning brands over long periods of time through the mass media. If you walk onto a used car lot, direct persuasion is likely to follow. The salesperson is interested in your driving out in a different car, not positioning the product in your mind. Within the context of mass media advertising, however, consumers don't have the opportunity to buy the brand *at that moment*, so positioning becomes paramount (although it should be noted that the internet is changing this).

Marketers have developed a number of techniques for positioning brands without drawing attention to the fact that they are positioning brands (Sutherland & Sylvester, 2000). Like magicians using slight-of-hand, a common way to accomplish this is with some kind of distraction. Because cognitive defenses are activated with persuasion awareness, advertisers often seek to divert audience attention so that persuasion information is accepted with less resistance and scrutiny. Four common techniques are humor, curiosity, narrative and metaphor.

Because repetition is critical for recall, advertisers like to hammer home their key selling points. This approach was made famous by Rosser Reeves of the Ted Bates Agency in the form of the "unique selling proposition." This approach, which advocates picking one aspect of brand superiority and reinforcing it over and over (and over), is still a staple in advertising and marketing texts. More empowered modern audiences (with remote controls and mice), however, can tune out when presented with annoyingly repetitive claims. Humor is a simple and effective way to distract people from the fact that they are being told—*over and over again*—that Geico could save them 15% or more on auto insurance (or that Miller Lite tastes great and is less filling, etc.). When the audience is being bemused by the clever ad, the brand positioning information is registering with little cognitive resistance.

Piquing people's curiosity is another tried-and-true indirect technique. Although curiosity is generally characterized as an aversive state (people don't like to have gaps in their knowledge), satisfying that curiosity (closing the knowledge gap) is enjoyable (Lowenstein, 1994). This gives curiosity the power to promote two things advertisers want to have happen—information seeking (to close the knowledge gap) and enjoyment (when the gap is closed). Bill Bernbach, the "Ad Man of the Century" according to *Ad Age,* and a key force in the "creative revolution" of the 1960s, was a master of this technique, which he called "effective surprise." This approach usually starts with some kind of unexpected or incongruous text or image ("incongruity resolution" is another label for it). Audiences are drawn in to seek out information in the ad that will explain what is going on. If well done, the curiosity-satisfying information includes brand positioning (this is the "effective" part of Bernbach's effective surprise). This technique remains one of the dominant approaches to advertising creative today.

Resistance to persuasion can also be lowered by engaging the audience in a story, a process often referred to as narrative transportation (Escalas, 2007; Green & Brock, 2000; Wang, 2006). David Ogilvy is often credited with "discovering" the power of stories in advertising, and narrative or "slice of life" appeals are quite common in commercial messaging. As audiences are engrossed in the mini-narrative in the message, they don't think about the fact that the advertiser is also conveying brand positioning information.

Finally, metaphors are often employed to quickly position brands and indirectly convey information in a way that "flies under the radar" of audience awareness (Coleman & Ritchie, 2011). Metaphors are powerful linguistic tools that activate broad sets of associations and meanings of which audiences are only partially cognitively aware (Lakoff & Johnson, 1999).

The indirect advertising techniques described above work to impede the audience's motivation to defend against persuasive information but not their ability. Therefore, most people do not consider the techniques themselves as highly problematic. In addition, advertising messages that use these techniques usually include some level of information or argumentation that can be independently and objectively verified in the context of established advertising ethical guidelines. The main advantage of using indirect approaches for advertisers is that over time more persuasive information makes it into the minds of consumers than would otherwise be the case. Therefore, when it comes time to make a brand choice, or to come up with a list of brands from which to choose, more brand positioning information will be accessible (Sutherland & Sylvester, 2000). These techniques do, however, diminish individual autonomy to the extent that they circumvent or suppress cognitive persuasion defense and can therefore be considered potentially ethically problematic.

Other non-conscious advertising techniques go further. These are the indirect image ads and values appeals described previously. These types of appeals often drop the delivery of linguistic information altogether and simply present a world in which the brand has a clear set of social-identity and affective associations. There is no truth or falsity to evaluate in these messages—they just are—and consumers have few if any cognitive defenses for them (Postman, 1986). There is mounting empirical evidence that these types of affective associations influence consumer attitudes and behaviors more strongly than consciously mediated persuasive information (Nairn & Fine, 2008).

Taken together, indirect non-conscious advertising techniques and the use of non-conscious imagery represent a spectrum of approaches to persuasion. At one end of this spectrum are completely conscious/informational appeals. Such messages do not in any way seek to impede persuasion awareness, and all information is presented in clear, verifiable linguistic terms. At the other end of the spectrum are completely non-conscious/values appeals. These ads associate the brand with a desirable terminal or instrumental value through some combination of imagery, narrative and/or metaphor, with no verifiable information explicitly provided. As noted above, most advertising uses a mix of these techniques. Subjectively placing an ad along this conscious/non-conscious spectrum, based on the information, techniques and appeals employed, is an important step in its ethical evaluation.

Ethics and the Non-Conscious

The moral philosophical underpinnings of most modern ethical evaluative frameworks are in the traditions of deontology, teleology and consequentialism. These traditions can be viewed as emphasizing the motives, actions and outcomes of message sponsors, messages and their effects (MacKinnon, 2004). Traditional moral philosophy is largely premised on the idea that people have a higher faculty that allows them to think independently and abstractly and which makes possible reasoned conclusions as to the morality of human actions. These assumptions have been criticized by cognitive scientists who argue that the ability for abstract moral reasoning is bound by human subjectivity in general and the metaphorical limits of language in particular (Lakoff & Johnson, 1999). Recent work in social psychology supports the notion that a good deal of ethics is based not on reason, but intuition. Over time collective intuitions build up into cultural virtues and accepted codes of behavior (Haidt & Joseph, 2004).

These efforts to understand ethics in the context of modern cognitive science and social psychology illustrate the importance of non-conscious factors in understanding behavior. They point out not the shortcomings of traditional moral philosophy but its limitations. If the non-conscious is essential to understanding ethics, ethical frameworks should appreciate the importance of non-conscious persuasion in commercial messages.

The Ethics of Non-Conscious Persuasion in Advertising

Arrington (1982), in an influential essay on this topic, argued that indirect information transfer in advertising is not unethical. Building on the central importance of personal autonomy in ethical communications, he concludes that "while advertising may violate my autonomy by leading me to act on desires which are not truly mine, this seems to be the exceptional case" (p. 7). He tackles the difficulty of extending the concept of autonomy to non-conscious desires with a market-oriented approach. While impulse purchases resulting from irrational desires stirred by image advertising can be seen as suppressing individual autonomy, Arrington argues this subjective effect is unlikely to have long-term consequences. Ultimately, individual autonomy is restored when the consumer has the opportunity to reflect on his/her satisfaction with the consumption of the product or brand. If the consumer's subjective desire is not fulfilled, they will not purchase it again in the future. If an individual has an opportunity to reflect on a second-order desire to be satisfied with their consumption choices, an indirect manipulation or activation of the first-order desire (for happiness, attractiveness, etc.) is not ethically consequential.

Santilli (1983) reaches the opposite conclusion: all advertising that operates by stimulating non-conscious desires is immoral. In order to support this conclusion, Santilli moves from the analysis of any one particular advertisement or advertising technique to the aggregate effects of all advertising on audiences: "Taken as a general, ongoing practice, addressed not to isolated individuals but to society as a whole, advertising which concentrates on the cultivation of wants through irrational means fails to meet the criterion of utility for it lessens the well-being of society by undermining the cognitive means for understanding what our real needs are" (p. 29).

The different conclusions reached by Arrington and Santilli are illustrative of an ongoing division in the literature. Those who conclude that advertising suppresses autonomy in a meaningful way tend to focus on the societal level (Klein, 2009; Lippke, 1989; Sneddon, 2001; Waide, 1987), while those who tend to defend advertising practices analyze the potential effects of individual ads on individuals (Bishop, 2000; Cunningham, 2003). Viewed from this perspective, it is possible to support both conclusions. Like junk food, consuming one ad may not have any direct negative effects, but making ads a permanent part of the information diet is likely to have serious long-term implications. As Cunningham (2003) concludes: "As such, advertising, though it may in other ways detrimentally affect society and consumers' well-being, does not violate our autonomy" (p. 236).

In an effort to bridge this gap, Lippke (1989) explores the concepts of dispositional and global autonomy. Because individual or dispositional autonomy is subjective, it is difficult if not impossible to define universal rules relative to its operation. One man's want is another man's need, for example, and individuals have different interacting levels of desire (Waide, 1987). Global autonomy, however, depends on certain objective social conditions, including the ability to rationally debate and scrutinize the structure of the political and economic institutions under which one lives. Lippke (1989) argues that globally autonomous individuals must "subject claims they are confronted with and norms others urge on them to rational scrutiny" (p. 40), especially relative to institutions that are "humanly alterable" (p. 40).

This line of reasoning is reflective of the tenets of public sphere theory (Habermas, 1989). In this analytical framework, rational debate is seen as critical to legitimate public policy development, and advertising is considered a corrupting influence. Here, the underlying universal value of the right to legitimate social organizations and self-determination is analogous to the idea of global autonomy. Although Habermas does not refer to any one type of advertising (conscious/informational versus non-conscious/values advertising), his argument seems to emphasize the social perils of indirect image appeals: "For the criteria of rationality are completely lacking in a consensus created by sophisticated opinion-molding services under the aegis of a sham public interest. Intelligent criticism of publicly discussed affairs gives way before a mood of conformity with publicly presented persons or personifications; consent coincides with good will evoked by publicity" (p. 85).

This review suggests that while individual indirect image advertisements are not necessarily unethical, global autonomy and the functioning of the public sphere require rational debate on issues of public concern. To the degree that an individual indirect image appeal diminishes or impedes such rational debate on a public issue, it can be seen as ethically problematic. The ethics of non-conscious advertising, therefore, establish another dimension on which such advertising must be evaluated: the degree to which the issue or subject of the advertising is an area of public debate. This consideration also illustrates the importance of evaluating image ads in terms of the values they directly or indirectly evoke. These values, not the images, narratives or metaphors used to evoke them, are the subject of the advertising.

Establishing the Publicality of an Issue

Some issue areas are clearly not a matter of public debate, whereas others clearly represent "hot" topics of public concern. What constitutes a public issue is different in different cultures and changes over time. The work of Herbert Blumer has been widely used to support the objective identification of public opinion and public issues (Blumer, 1951; Blumer, 1971). Blumer defined "publics" by the issues/topics they were actively engaging in the public sphere. As issues become areas of public debate, publics emerge on either side to play a role in the discussion. As issues fade, so do the active publics which supported and created them. The public sphere, in this context, comprises the media and strategic communications campaigns that use other elements of the modern media mix. A public issue, by this objective definition, is one that is being actively engaged by opposing publics in the mass media and/or through active strategic communications campaigns.

Based on the foregoing, the degree to which an issue area is the subject of public debate is another dimension on which advertising must be ethically evaluated. This dimension can be combined with the direct/indirect spectrum of advertising techniques and appeals described earlier to create a useful evaluative map of advertising ethicality.

Perceptual Mapping and Ethical Evaluation

Perceptual mapping is a method used to spatially graph and compare cases on two orthogonal dimensions. This approach has been standard practice among marketers for decades (Moriarty, Mitchell, & Wells, 2009). In its simplest form, the method plots cases on the two dimensions of interest, resulting in four quadrants that represent meaningful groupings of cases that aid in the interpretation of individual cases. A common example from marketing would be to plot different automobile brands on a horizontal axis representing the continuum of "conservative" (left endpoint) to "sporty" (right endpoint), and on a vertical axis representing the continuum of "practical" (bottom endpoint) to "luxury" (upper endpoint) (Moriarty et al., 2009). In this example, the Porsche and BMW brands would be located in the upper right "sporty/luxury" quadrant. Of course, marketers have built on this basic model over the years to add multiple vectors and levels, which are graphed in three dimensions and analyzed with advanced statistical techniques such as factor analysis, discriminant analysis and multidimensional scaling (Kohli & Leuthesser, 1993).

Perceptual maps can be used to measure consumer attitudes toward existing brands or to construct ideal brands based on consumer preferences. Non-empirical evaluative maps can also be employed to analyze abstract ethical relationships. Bendinger et al. (2004), working with ethical dimensions explored by Jaska and Pritchard (1994), used an evaluative map to classify advertising cases on the ethical dimensions of target susceptibility (i.e., children and old people) and potential product harm. In this construction, message cases fixed in the upper right quadrant (high susceptibility target/high harm product) are considered the most ethically problematic, whereas message cases in the lower left quadrant (low susceptibility target/low harm product) are seen as the least ethically problematic.

A similar evaluative map can be employed to analyze the ethicality of non-conscious advertising. For this analysis, the two orthogonal ethical dimensions are level of non-conscious appeal (vertical axis) and degree of public issue (horizontal axis). The bottom end-point of the vertical axis would represent purely conscious/informational appeals and the upper end-point would represent purely non-conscious/values appeals. The left end-point of the horizontal axis would represent completely non-public issues, and the right end-point of the horizontal axis would represent hotly contested public issues (see Chart 1). In this ethical framework, ads that are completely conscious/informational are always ethical (the bottom axis in the chart). Advertising that focuses on completely non-public issues are also ethical, regardless of the techniques and appeals used to deliver the persuasive information to audiences (the left-most axis of the chart). The combination, however, of non-conscious/values appeals and public issues becomes the most ethically problematic. Advertising that focuses on a highly public issue and uses some level of non-conscious appeal would be considered potentially problematic as would advertising that uses a highly non-conscious approach and is focused on an issue of some level of public concern.

Chart 1: Advertising Ethicality Evaluative Map (AEEM)

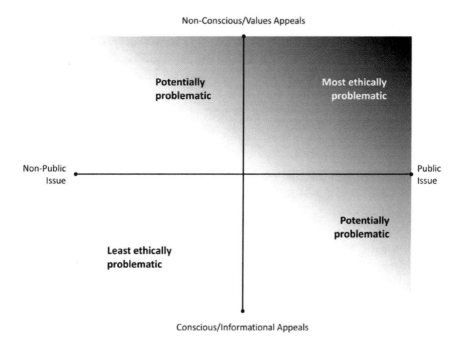

Non-Conscious/Values Appeals

Potentially
problematic

Most ethically
problematic

Non-Public
Issue

Public
Issue

Potentially
problematic

Least ethically
problematic

Conscious/Informational Appeals

An Extended Framework for the
Ethical Evaluation of Advertising

This chapter builds on prior research in advertising and communications ethics to develop an extended framework for the ethical evaluation of advertising. The approach advocated here builds on a number of premises that are supported in previous literature and past research.

- Individual ads can be subjectively evaluated as to the nature and level of non-conscious techniques and image/value appeals employed.
- The issue area that is the primary subject of individual ads can be objectively evaluated in terms of the degree to which it represents a public issue.

- Issues of public concern should be presented in a way that allows audiences to engage in rational debate and evaluation of the explicit and implicit claims being made.

This framework extends past theorizing by conceiving of the continuum of indirect approaches to advertising as a continuous variable and combining this dimension of advertising ethicality with the degree to which the issue at hand is in the domain of public concern. The result of this extension is the development of an advertising ethicality evaluative map (AEEM) that spatially situates individual advertising cases in a way that aids in moral interpretation (AEEM analysis).

This approach allows for more subtle distinctions between messages, takes into account interactions between these two key dimensions of ethicality, and can be seen as reconciling divergent views on advertising morality. For example, Santilli (1983) concludes that while informational advertising is always ethical (presumably truthful informational advertising), persuasive (non-conscious) techniques and appeals are always unethical. This conflicts with Arrington's (1982) view that such image appeals do not violate autonomy in a meaningful way. The advertising ethicality evaluative map indicates that (truthful) conscious/informational advertising is always ethical, but the ethicality of non-conscious/values appeals depends on the publicality of the central issue of the message.

The Process of Evaluating Advertising Ethicality

The initial step in the ethical evaluation of advertising is to closely review the message. Does the advertisement employ some level of non-conscious persuasion? Does the advertisement associate or evoke some abstract but desirable terminal or instrumental value? What amount of concrete information or argumentation does the message directly or indirectly convey? All these factors should be considered as a whole when subjectively placing an advertisement in terms of non-conscious persuasion level. A message that uses imagery and emotional associations may also provide some level of concrete information that the audience can consider in a rational way. Other messages may not use affective imagery but employ other techniques that diminish or work to circumvent cognitive persuasion awareness. Specific guidelines can be developed for the subjective interpretations inherent in each of these steps, just as they have been for the ethical evaluation of informational advertising. For example, precise definitions have evolved to determine if an informational advertisement is deceptive.

To determine if the subject of an advertisement is a current issue of public concern, one can simply ask the question "Is there currently a public debate over how society should achieve [value]?" As the preceding definition suggests, this will be time dependent. For example, how society should achieve national security may be a fierce public issue during times of war but not achieve the level of a public issue during times of extended peace. While it might not generate much notice if a marketer were to somehow associate a brand with national security in an advertisement during peacetime, it may be quite controversial to do so in the midst of a hot war.

Some extreme examples illustrate the basic functionality of AEEM analysis. Perhaps the most potent recent example of a public issue would be national security in the wake of the September 11 attacks. Clearly, any advertisement evoking such a sensitive public issue using a non-conscious/values appeal would be considered ethically problematic (for cynically capitalizing on a tragedy). At the other end of the spectrum would be a non-controversial and therefore non-public issue such as convenience. Clearly, an advertisement using a purely conscious/informational approach to promote a brand as delivering consumer convenience would be not be considered ethically problematic (assuming it is truthful).

The two environmental advertisements featured at the beginning of this chapter help illustrate how AEEM analysis informs more subtle distinctions. For both ads, the central value being evoked is environmentalism (or sustainability to use a more modern term). Based on the objective criteria described above, this is clearly a public issue. There are active publics on both sides of the debate who are competing in the public marketplace of ideas. There are oppositional voices in the mass media, and strategic communicators from both sides of the issue are sponsoring public information campaigns to support their points of view. Both advertisements, therefore, would be placed near the public issue (right end-point) on the publicality dimension (Chart 2).

The messages are somewhat different, however, when it comes to the level of non-conscious appeal. The BP ad evokes a "green world" and associates its brand with this terminal value but provides no direct or indirect information as to how the brand is instrumental in achieving the value. Does BP use wind turbines to power its gas stations? Is BP gasoline produced and marketed using more environmental and sustainable practices (compared to the competition)? No such syllogistic claims or connections are directly or indirectly presented. This ad, therefore, would be placed near the top end-point in terms of the non-conscious/values appeals employed.

The Nissan Leaf ad also uses non-conscious persuasive techniques and values imagery. The polar bear narrative engages the audience and elicits positive affect, diminishing cognitive scrutiny of the main message claim: that the 100% electric Nissan Leaf is better for the environment than other brands of automobile. The ad does, however, make a clear claim that the audience can rationally evaluate. The reduced emissions from an electric car, and the degree to which these reduced emissions will reduce greenhouse gases, can be empirically verified and rationally considered. This concrete (if indirect) verifiable claim places the ad more toward the conscious/informational end-point on this dimension. Subjectively placing the ad near the mid-point on this dimension, or just below the mid-point, seems appropriate.

The relative placement of these two messages on the AEEM is illustrated in Chart 2. The lack of any verifiable information makes the BP ad most ethically problematic. The Leaf ad, because it presents some level of rational argument, is more accurately labeled potentially problematic.

Chart 2: AEEM Analysis of Two Green Ads

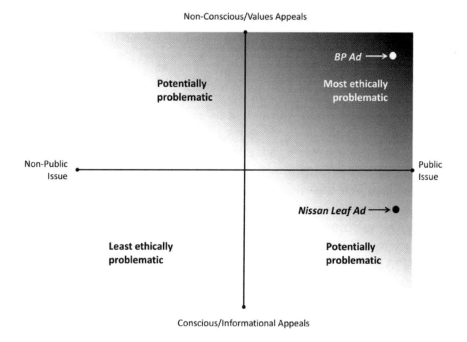

Conclusion

AEEM analysis provides regulators, critics, audiences and advertisers themselves a useful tool for morally classifying approaches to persuasion that have heretofore avoided systematic scrutiny. As an *extended* framework for the ethical evaluation of advertising, it should be reiterated that the concrete, verifiable information and argumentation directly or indirectly presented in the advertising must still be subjected to the existing, well-developed ethical guidelines discussed previously. AEEM analysis adds another useful layer of evaluation after an ad's verifiable claims (if any) have been evaluated for truthfulness and accuracy.

Advertising messages that are factually accurate but which also convey significant non-conscious associations relative to an important public issue are considered potentially ethically problematic. Ads that rely exclusively on non-conscious/values appeals relative to issues in the domain of public concern are considered ethically problematic. In this way, observers and audiences can hold marketers and strategic communicators to a more rigorous moral communicative standard. Advertisers will not be able to circumvent critical ethical scrutiny by employing non-conscious appeals. At the same time, not all non-conscious techniques are deprecated with a blanket condemnation.

By treating two key dimensions of advertising ethicality as continuous variables, this framework allows for closer comparison of individual messages. Rather than group all ads, or all ads of a certain type, as ethical (or unethical), this approach takes into consideration the interaction of these ethical dimensions, facilitating more subtle distinctions. A detailed taxonomy of non-conscious advertising techniques is presented, along with an objective method for the identification of public issues. Within the basic framework of AEEM analysis, more detailed and specific methods of subjective evaluation can be developed. By aiding in the interpretation of advertising ethicality, this approach can help protect the virtues of global autonomy and public sphere debate, to which unchecked advertising is a natural threat.

Specifically, this approach provides ethical guidance for nonprofit and pro-social organizations that struggle with the implication of using modern indirect advertising techniques to promote their causes. By considering the publicality of the issue or value being evoked with the degree to which a message operates nonconsciously or indirectly, it is possible to distinguish if and when the use of such techniques is ethically problematic. Marketers and pro-social advocates can be held to the same standards of ethical communication yet remain free to use modern persuasive approaches when fair and appropriate.

References

Arrington, R. L. (1982). Advertising and behavior control. *Journal of Business Ethics, 1*(1), 3–12.

Belch, G., & Belch, M. (2009). *Advertising and promotion* (Eighth ed.). New York: McGraw-Hill Irwin.

Bendinger, B., Maxwell, A., Barnes, B., Tucker, E., McGann, A., Gustafson, R., et al. (2004). *Advertising & the business of brands*. Chicago, IL: The Copy Workshop.

Bishop, J. D. (2000). Is self-identity image advertising ethical? *Business Ethics Quarterly, 10*(2), 371–398.

Blumer, H. (1951). Collective behavior. In A. M. Lee (Ed.), *New outline of the principles of sociology* (pp. 166–222). New York: Barnes & Noble.

Blumer, H. (1971). Social problems as collective behavior. *Social Problems, 18*(3), 298–306.

Chaiken, S., Libermann, A., & Eagly, A. H. (1989). Heuristic and systematic information processing within and beyond the persuasion context. In J. S. Uleman, & J. A. Bargh (Eds.), *Unintended thought* (pp. 212–252). New York: Guilford.

Coleman, C., & Ritchie, D. L. (2011). Examining metaphors in biopolitical discourse. *99th Annual Conference of the Association of Educators in Journalism & Mass Communication*, St. Louis, MO.

Crisp, R. (1987). Persuasive advertising, autonomy, and the creation of desire. *Journal of Business Ethics, 6*(5), 413–418.

Cunningham, A. (2003). Autonomous consumption: Buying into the ideology of capitalism. *Journal of Business Ethics, 48*(3), 229–236.

Escalas, J. E. (2007). Self-referencing and persuasion: Narrative transportation versus analytical elaboration. *Journal of Consumer Research, 33*(4), 421-429.

Fishbein, M., & Ajzen, I. (1975). *Belief, attitude, intention, and behavior: An introduction to theory and research*. Reading, MA: Addison-Wesley.

Green, M. C., & Brock, T. C. (2000). The role of transportation in the persuasiveness of public narratives. *Journal of Personality and Social Psychology, 79*(5), 701–721.

Habermas, J. (1989). *The structural transformation of the public sphere*. Cambridge, MA: The MIT Press.

Haidt, J., & Joseph, C. (2004). Intuitive ethics: How innately prepared intuitions generate culturally variable virtues. *Daedlus, 133*(4), 55–66.

Jaska, J. A., & Pritchard, M. S. (1994). *Communication ethics methods of analysis*. Belmont, CA: Wadsworth Publishing Company.

Klein, N. (2009). *No logo: No space, no choice, no jobs*. New York: Picador.

Kohli, C. S., & Leuthesser, L. (1993). Product positioning: A comparison of perceptual mapping techniques. *The Journal of Product and Brand Management, 2*(4), 10.

Lakoff, G., & Johnson, M. (1999). *Philosophy in the flesh.* New York: Basic Books.

Lippke, R. L. (1989). Advertising and the social conditions of autonomy. *Business & Professional Ethics Journal, 8*(4), 35.

Lowenstein, G. (1994). The psychology of curiosity: A review and reinterpretation. *Psychological Bulletin, 116*(1), 75–98.

MacKinnon, B. (2004). *Ethics: Theory and contemporary issues.* Belmont, CA: Thompson Wadsworth.

Marlin, R. (2003). *Propaganda and the ethics of persuasion.* Ontario: Broadview Press.

Moriarty, S., Mitchell, N., & Wells, W. (2009). *Advertising principles & practices* (Eighth ed.). Upper Saddle River, NJ: Pearson.

Nairn, A., & Fine, C. (2008). Who's messing with my mind? *International Journal of Advertising, 27*(3), 447–470.

Ogilvy, D. (1985). *Ogilvy on advertising.* New York: Vintage Books.

Postman, N. (1986). *Amusing ourselves to death.* New York: Penguin Books.

Rokeach, M. (1968). *Beliefs, attitudes, and values.* San Francisco: Jossey-Bass.

Santilli, P. C. (1983). The informative and persuasive functions of advertising: A moral appraisal. *Journal of Business Ethics, 2*(1), 27–33.

Sneddon, A. (2001). Advertising and deep autonomy. *Journal of Business Ethics, 33*(1), 15–28.

Spence, E., & Van Heekeren, B. (2005). *Advertising ethics.* Upper Saddle River, NJ: Pearson Education, Inc.

Sutherland, M., & Sylvester, A. K. (2000). *Advertising and the mind of the consumer.* St. Leonards, NSW, Australia: Allen & Unwin.

Waide, J. (1987). The making of self and world in advertising. *Journal of Business Ethics, 6*(2), 73–79.

Wang, J. (2006). Lost in the story: Factors that affect narrative transportation and advertising. *Advances in Consumer Research, 33*(1), 406–408.

Finding Common Cause

Tom Crompton

This is an important and stimulating collection of essays that takes "Weathercocks and Signposts" as a springboard for deeper reflection on the strategies that the environment movement is currently pursuing—and on the crucial question of how to bridge the chasm (a widening chasm, it seems) between the scale of environmental problems that the science lays before us and the scale of response that is currently mustered.

On many estimates, in 2008 when WWF-UK published "Weathercocks," public concern about climate change was peaking: much has changed in the intervening years, and environmental concern has proved fickle in the face of economic recession (Scruggs and Benegal, 2012). Now, in the UK at least, government is openly stalling on some existing international environmental commitments on grounds of economic cost (Vidal, 2012). Clearly, this doesn't augur well for the necessary international momentum to commit to far more ambitious action on environmental problems such as climate change.

Should the fragile nature of public concern about the environment be seen to illustrate the futility of building public commitment to tackling *environmental* problems upon *environmental* concern? Is the environment movement still too hesitant to cast appeals to public environmental concern to one side, and focus exclusively on the economic opportunities arising in the course of taking action on (at least some) of these problems? A great deal of environmental discourse—particularly, it seems, in the US—argues that both these questions are to be answered "yes."

Or was the strategy to marry economic opportunity with environmental action ill-conceived at the outset? Was it fundamentally short-sighted to insist that thorough-going responses to environmental problems could be anticipated to emerge tangentially, while economic concerns continued to drive political process and—to a large extent—public concern? Does the recent erosion of public commitment to the environment rather underscore the need to elevate public concern about social justice and environmental degradation above public concern about today's politically salient economic challenges (which are, by comparison, slight)?

These are surely among the most pressing questions confronting environmentalism today. They are questions which "Weathercocks," and many of the contributions in this volume, help to explore. For example, Nisbet and colleagues argue that campaigns in the US which led on the economic and employment opportunities offered by investment in renewables failed to galvanize public support—particularly in the context of recession. "The emphasis on economic benefits in the context of the recession," they write, "...turned the debate into 'some economic benefits' as claimed by greens versus 'dramatic economic costs' as claimed by opponents, a balance that given the economic context favoured the opposition" (this volume, p. 20).

This may always be the case. "Weathercocks" pointed towards a body of work in social psychology that highlights a fundamental tension between economic concern and environment concern, operating at the level of values. It seems that the greater the importance that an individual—or, indeed, a culture—attaches to values of wealth and financial success, the less importance he or she is likely to place upon social justice or protecting the environment.

Yet, still, this doesn't matter if there is a near complete convergence of economic opportunity with environmental imperative—such that environmental problems are addressed "in passing" in pursuit of economic goals. The environment needn't then especially concern us—we can rely upon economic interests to get the environmental job done. But if this convergence is only partial, it seems that privileging economic concerns above environmental concerns risks eroding environmental concern with potentially devastating consequences, fundamentally undermining public appetite for systemic action on environmental problems.

So, is the convergence of economic opportunity with environmental imperative all but complete? "Weathercocks" marshalled evidence that this is far from the case. Indeed, the report helped to catalogue the reasons why we currently wildly over-estimate the extent of this convergence. The problems arise from the unlikelihood that ambitious environmental action will be motivated through the extension of small steps, taken in pursuit of financial savings (Crompton, 2008; Thøgersen & Crompton, 2009) or from the problems of rebound (Crompton, 2008). But they also arise from a range of more structural constraints—inertia to changing institutions and incentives that benefit powerful vested-interest groups and which promote environmentally damaging activities. How do we foresee the emergence of irresistible public demand to travel less, or to invest in public transport, or to invest in renewables—but not carbon-intensive domestic energy sources—or for governments to develop robust international agreements on transboundary environmental problems (such as climate change or water shortage)?

If the scale of environmental problems with which we are currently confronted will not be adequately addressed by relying upon the business opportunities that these present, then it seems we must hope for the emergence of *very* much stronger political momentum for change. Such change will include new incentives and regulations for business but also public investment in new institutions and infrastructure. How can we envisage this political momentum emerging? Certainly, new and bold political leadership will be very important. But it is difficult to foresee such leadership emerging and flourishing in the absence of far stronger public demand for environmental action. And such demand will be importantly premised upon what, as a society, we take to be important—our values. We must ask: what might the environment movement do to help strengthen those values which underpin public acceptance of—and appetite for—far more ambitious and, at times expensive, action on environmental problems?

Five years ago, in the UK, there was little serious challenge among mainstream environmental organizations to what "Weathercocks" characterized, and critiqued, as the marketing approach: a set of techniques borrowed from commercial marketing and relying upon the segmentation of audiences by some set of characteristics (sometimes dominant audience values) and then framing exhortations to adopt pro-environmental behaviors through appeal to these characteristics. In particular, "Weathercocks" took aim at the insistence that some audience segments can only be meaningfully engaged through appeal to self-interest. As one environmental consultant wrote, in a 2006 paper summarizing principles for environmental communication drawn from discussion with many environmental campaigners and communicators:

> An accurate basic assumption might be that most people are essentially selfish....Any benefits from environmental behavior, and there should be benefits from every environmental behavior, must be tangible, immediate and specific to the person carrying out the behavior. Benefits at the society level are unlikely to be a significant driver of change. (Hounsham, 2006, p. 139)

Such perspectives are psychologically naïve. But, because they appeal to socially and environmentally unhelpful values, they are also likely to be dangerous.

"Weathercocks" helped to widen discussion on the dangers associated with appealing to self-interest as a means of motivating increased uptake of behavior with environmental benefits. Reassured by the widespread appetite for this discussion—even among larger and therefore, perhaps, more risk-averse environmental organizations—and emboldened by the support of academics who stepped forward from a range of disciplines, WWF-UK widened this discussion further with the subsequent publication of *Meeting Environmental Challenges: The Role of Human Identity* (Crompton & Kasser, 2009), *Common Cause: The Case for Working with Our Cultural Values* (Crompton, 2010), and *The Common Cause Handbook* (Holmes et al., 2011).

As these studies highlight, substantial cross-cultural research has identified a limited set of values and life goals that consistently emerge across nations. The organization of these values and goals is remarkably consistent across cultures. Among these is a set of values and goals focused on wealth, rewards, achievement and status.

For example, Shalom Schwartz and colleagues (1992, 2006) have identified the existence of two sets of *self-enhancement* values that consistently emerge across 70 nations as fundamental and coherent aspects of people's value systems. Schwartz calls these values for *power*, or the desire to dominate people and resources, and for *achievement*, or the desire to demonstrate one's success relative to others. Other work by Grouzet and colleagues (2005) across fifteen nations has documented the cross-cultural emergence of a related set of life goals, labelled *extrinsic*. These goals, which are focused on the attainment of external rewards and social praise, include specific aspirations for financial success, popularity and having a socially desirable image.

Quantitative empirical studies document that people who strongly endorse such self-enhancement values and extrinsic goals express lower concern about both social and environmental problems and are less likely to engage in behavior that would help to ameliorate these problems (see Crompton, 2010 or Holmes et al., 2011, for review). To the extent that an individual holds self-enhancement values and extrinsic goals to be important, they are likely to attach less importance to self-transcendence values and intrinsic goals. Yet self-transcendence values and intrinsic goals are associated with greater concern about both social and environmental problems and greater likelihood to engage in behavior that would help to ameliorate these problems. This oppositional relationship between self-enhancement and self-transcendence values, and extrinsic and intrinsic goals, is also found in a dynamic way: priming self-enhancement values (for example, subtly drawing a person's attention to money) diminishes the importance that he or she places on self-transcendence values and reduces the frequency with which he or she expresses pro-social or pro-environmental attitudes. It also reduces the frequency with which he or she engages in pro-social or pro-environmental behaviors.

Consistent with these empirical findings, it is also found that framing environmental communications and campaigns in terms of self-enhancement values or extrinsic goals results in lower motivation to engage in pro-environmental behavior as compared to framing these communications and campaigns in terms of intrinsic values.

Greg Maio and colleagues at Cardiff University, for example, found that when participants in a study were presented with information about car sharing schemes *that included reference to the money that might be saved by joining such a scheme* they were less likely to recycle scrap paper (Evans et al., under review). It seems that drawing attention to the possibility of saving money activated self-enhancement values and reduced the importance that a person attaches to self-transcendence values, thereby eroding motivation to adopt behaviours that are environmentally beneficial. Salmon and Fernandez (this volume) draw attention to the importance of considering unintended consequences in designing interventions to address environmental problems. Among the key factors leading to unintended consequences they list "firmly entrenched cultural values." These, they suggest, "may strongly influence and delimit the range of policy options [and, one might add, other interventions] that can be brought to bear on a social problem, thereby resulting in a less-than-optimal 'satisficing' strategy that appeases stakeholders but fails to address root causes" (p 39).

Experiments such as that conducted by Maio and colleagues suggest that unintended consequences may arise where strategies aimed at increasing uptake of private-sphere energy-saving behavior (such as participating in a car-share scheme) are framed in terms of appeal to financial savings.

An understanding of values and their importance in motivating expressions of pro-social and pro-environmental concern also helps to explain the failure of the 'information deficit model' of environmental advocacy. Some environmental communicators and campaigners still labour under the conviction that most individuals will be motivated to express concern about environmental problems once confronted with the *facts* of these problems (Crompton, 2008). Yet there is copious evidence—both experimental and anecdotal—that facts alone usually present a poor means of motivating practical expressions of concern. As Dan Kahan at the Cultural Cognition Project at Yale Law School writes:

> The prevailing approach is still simply to flood the public with as much sound data as possible on the assumption that the truth is bound, eventually, to drown out its competitors. If, however, the truth carries implications that threaten people's cultural values, then...[confronting them with this data] is likely to harden their resistance and increase their willingness to support alternative arguments, no matter how lacking in evidence. (Kahan, 2010, p. 297)

It is crucial to differentiate between change models which rely narrowly upon the assumed persuasive effects of information and models which highlight the importance of promoting civic participation and self-governance while recognizing the importance of information as a necessary precondition. Of course, as Palenchar and Motta (this volume) stress in discussing "right to know" as a basis for civic participation in risk analysis, access to reliable information is crucially important. But the provision of such information, alone, is unlikely to motivate increased public deliberation in the absence of a culture which supports and promotes self-determination and concern about social justice—such that citizens themselves come to make demands for reliable information upon which to base their deliberations. Palenchar and Motta also acknowledge this, writing:

> The core value of right to know is about clarity of not only issues but also values that underlie and motivate behavior. (p. 94)

They also cite Heath and colleagues (2007), who make this point still more clearly:

> It is becoming increasingly clear that the main product of environmental and risk communication is not informed understanding as such, but the

quality of the social relationship it supports, becoming a tool for communicating values and identities as much as being about the awareness,
attitudes, and behaviors related to the risk itself (p. 46).

A culture which promotes civic participation will also promote—
and be promoted by—universalism and self-direction values (closely associated with the self-transcendence values and intrinsic goals
discussed above). Values of universalism and self-direction, if
strengthened, are likely to be associated with wider expressions of
social and environmental concern. So "Weathercocks" certainly did
not advocate that governments respond to environmental problems
by forcing through new policy in the face of public opposition: were
that politically possible (which it is usually not) it would be unlikely to build public appetite for further ambitious government interventions. So Gangadharbatla and Sheehan (this volume) are
absolutely right to argue that "it is imperative that environmental
activists work toward gathering more support for the cause rather
than force governmental regulations on an unwilling and an
unenthusiastic population." But they are mistaken to suggest that
"Weathercocks"argues otherwise. Nor, unfortunately, does the
study that they report here begin to address the deeper questions
of environmental strategy that "Weathercocks"sought to highlight.

"Weathercocks," and subsequent work developing these arguments, drew attention to the well-documented problem of rebound
(Crompton, 2008); profound uncertainty about whether appeals to
adopt particular pro-environmental behavior necessarily 'spillover'
into other—even more significant—pro-environmental behaviors
(Crompton, 2008; Thøgersen and Crompton, 2009); possible 'co-
lateral damage' associated with appeals to extrinsic goals (both
environmental messages tested by Gangadharbatla and Sheehan
in the research they report here appealed to extrinsic goals)
(Crompton & Kasser, 2009; Crompton, 2010); related problems of
promoting 'green consumerism' (Crompton, 2008; Crompton &
Kasser, 2009); problems of ensuring on-going commitment to pro-
environmental behavior (Crompton, 2008; Crompton, 2010); and,
most pressingly, the challenge of working to strengthen those cultural values which extensive research suggests underpin pro-

environmental attitudes and behaviors at a systemic level (Crompton & Kasser, 2009; Crompton, 2010). These are not concerns that Gangadharbatla and Sheehan (this volume) address in their study.

Nonetheless, Gangadharbatla and Sheehan are right to highlight the need for more work to explore the way in which the effectiveness of a pro-environmental communication in motivating pro-environmental attitudes or behaviors is likely to be moderated by the specific characteristics of the audience.

They suggest that level of environmental concern presents one audience characteristic that is likely to have a moderating effect. But there are others. Also of interest is the potential moderating influence of an audience's dominant values and life-goals. But we do not know how the motivational benefits of framing appeals to self-transcendence values are moderated by the values upon which an audience segment is disposed to place particular importance. More work is urgently needed here. We do know, however, that briefly priming self-transcendence values in study participants for whom self-enhancement values are of particular dispositional importance leads these participants to express significantly greater concern about a range of social and environmental issues, such as climate change (Chilton et al., 2011).

It is also the case that we need to develop new approaches to evaluating the full impacts of particular communications and campaigns—as these impacts extend beyond motivating the particular behavior that a campaign is designed to promote. Salmon and Fernandez (this volume) are clear on this point:

> Yet common to all types of unintended effects is the difficulty in assessing their validity. Unintended consequences usually cannot be measured or studied at the time at which the claim of their inevitability is made. Unintended consequences usually are the result of a complex chain of micro- and macro-level processes that do not lend themselves to clear-cut analysis or unambiguous causal linkages. (p. 54)

Of course, the difficulty of measuring the impacts of such unintended consequences cannot be allowed to detract from an appreciation of their importance in frustrating the emergence of wider commitment to ambitious environmental action.

How then can environmental communicators and campaigners respond to an understanding of values and how these work? I suggest that there are two important implications. The first is to ensure that, so far as possible, environmental communications and campaigns are designed to activate self-transcendence values and intrinsic goals. The second is to begin to identify and oppose those influences that currently serve to strengthen self-enhancement values and extrinsic goals in society, and to strengthen those influences that serve to promote self-transcendence values and intrinsic goals. This implies the need for environmental NGOs (and, indeed, NGOs working on social problems) to begin to work on issues that have hitherto been relatively overlooked by most NGOs—the cultural impacts of advertising, for example.

I want to focus first on the ways in which environmental organizations communicate on their core issues. Several contributions to this volume stress the influence that environmental organizations exert culturally through their existing communications. As Short (this volume) argues, all organizations have a profound collective impact in shaping cultural values:

> Organizations hold a privileged place in contemporary society. They typically are not part of the established political order, but hold great power in shaping how people construct meaning and respond to the world. (p. 61)

This is a perspective with which Palenchar and Motta (this volume) also agree:

> Our values, and our sense of who we are as individuals, organizations, and as a global society confronted with environmental challenges...are socially and symbolically constructed from a sense of those values of all organizations engaged in discourse around this topic. (p. 99)

Among these organizations, it may be that NGOs have a particularly important role, relative to the volume of communications that they produce, because of the relative trust that they command (Edelman, 2011). Sinclair and Miller (this volume) write of a particular type of communication—market advocacy messages:

> When perceptions of accountability and trustworthiness are strong, audience members are more likely to identify with the values in a marketplace advocacy message and less likely to critically evaluate the message. (p. 115)

This underscores the importance of examining the impacts of an organization's communications on those who see them. Bortree's work (this volume), reporting on an analysis of Twitter feeds generated by the Sierra Club, is one of few studies that have embarked upon analysis of a representative sample of NGO communications for the values that they are likely to activate.[1]

The way in which such communications are framed will be important in two ways. Firstly, it will be important in shaping the quality of relationship that an organization establishes with its supporters. Secondly, it will be important in shaping wider public attitudes to the environment (and, indeed, a wide range of other social concerns). Bortree (this volume) raises the important possibility that strategies adopted to deliver one of these outcomes need not be optimal in facilitating the other.

That is, some communication strategies designed to encourage direct support for an organization (e.g. financial support or volunteering) may not be particularly effective in helping to motivate wider environmental concern (e.g., commitment to simple domestic energy-efficiency behaviors or considering the environment in making decisions about how to vote—behaviors that are environmentally helpful but that do not entail commitment to give time or money to a particular environmental organization). Indeed, these two strategies may even conflict—particularly where, for example, fundraising efforts appeal to self-enhancement values and extrinsic goals.

---

[1] A group of a dozen conservation organizations in the UK have recently contracted social psychologists and linguists to analyze their entire communications output over a six-month period, examining the values that these communications are likely to activate. This is with a view to identifying—and in time deepening—the contribution that these organizations make to activating and strengthening self-transcendence values and intrinsic goals. This study will appear in early 2013, at www.valuesandframes.org

It is not clear that observed differences between strategies are driven by objective assessment of which approach best delivers particular outcomes. Differences could reflect a bias in untested assumptions made by NGO staff with different organizational roles. For example, staff in a supporter acquisition division are perhaps more likely to have backgrounds in marketing than staff in a policy division and may also be more likely to contract aspects of supporter acquisition to outside marketing agencies. If fundraising communications often appeal to self-enhancement values and extrinsic goals, this may reflect the bias of the people who produce them, rather than a strategic decision, on the part of an organization, to frame these communications in this way. Indeed, it is possible that fundraising communications that appeal to self-transcendence values and intrinsic goals would encourage greater supporter loyalty.

By way of example, consider a recent advertising campaign by a UK environmental organization to promote uptake of its corporate credit card scheme (under which a small proportion of the value of credit card purchases would be donated to this organization). The strapline ran: "Defend our world while you shop."

It is possible that a strategy such as this will:

• foster a more transactional and therefore less loyal relationship between supporters and the environmental organization
• reinforce a set of values which are antagonistic to a durable and ambitious commitment to action on environmental issues.

But of course, if the impact of this communication is to be measured solely in terms of the number of additional credit-card applications that the organization received, it is possible that this advertisement would be rated as being successful.

Such questions also arise even when focusing exclusively on the effectiveness of a communication in building public support for a particular campaign. In reflecting on her results, Bortree suggests that there may be instances where framing communications in terms of self-enhancement values or extrinsic goals makes sense for motivating particular issue-specific behaviors (for example, she suggests, in promoting concern for energy independence). But this raises a deeper set of questions about how the Sierra Club comes to find itself in a position where it Tweets on energy independence five times more frequently than greenhouse gas reduction (as reported by Bortree). This would seem to reflect acceptance of a dominant public discourse that promotes economic and security concern (about energy provision) above social and environmental concern (about the impacts of climate change). Perhaps, here, the Sierra Club was responding to what it saw as a political reality: "energy provision" was seen as an issue with which it must, necessarily, work. But the 'energy independence' narrative is one that the environment movement had a hand in creating and supporting, and it is one that probably has very damaging environmental implications.

Some of these probable implications are specific to the energy debate. Others are general to environmental concern. The specific implications are that public demand for energy independence will often be articulated as support for carbon-intensive energy sources such as coal, oil-shales, and oil, where these can be secured domestically. The general implications are that this narrative serves to further embed the importance of a set of values—here economic and security-related values—which, as we have seen, are likely to be inimical to the emergence of stronger public support for action on a wide range of social and environmental issues.

Questions about how best to frame communications—whether these are designed to recruit new supporters or to encourage wider expressions of environmental concern—are crucially important and demand far more careful reflection.

Recall that I suggest there are two important implications for environmental organizations arising from an understanding of cultural values. The first, which I've discussed, is to ensure that so far as possible all environmental communications and campaigns are designed to activate self-transcendence values and intrinsic goals. The second, to which I now turn, is to find common cause in working at a more systemic level to promote self-transcendence values and intrinsic goals in society—and, conversely, to work to reduce the importance attached to self-enhancement values and extrinsic goals. This second implication relates not just to the way an organization campaigns but also to the *issues* upon which an organization focuses these campaigns.

It seems that activating one self-transcendence value 'bleeds-over' into the activation of other self-transcendence values. So, for example, activating values related to social justice and benevolence leads to an increase in expressions of environmental concern (Sheldon et al., 2011). This finding underscores the likely benefits of framing communications about climate change in terms of both environmental and humanitarian impacts (cf. Nisbet et al., this volume).

"Weathercocks" unduly stressed the opportunities for campaigners and communicators to appeal to *environmental* imperatives for adopting pro-environmental behavior. It also failed to underscore the importance of meeting people where they are—understanding an audience segment and communicating in a way to which this segment will be open. More recently, I have argued for the importance of both (i) recognising the potential significance of non-environmental self-transcendence values and intrinsic goals in motivating behavior with environmental benefit, and (ii) engaging particular audiences through their specific interests or concerns *where these are associated with self-transcendence values or intrinsic goals*, while *avoiding* appealing to self-enhancement values or extrinsic goals. This is what Nisbet et al. (this volume, p. 31) seem to advocate when they call for greater "focus on framing the relevance of climate change and related challenges in a manner that activates a broader palette of moral intuitions, tailoring these appeals to specific segments of the American public." I agree with Nisbet and colleagues that more research is needed here. Some such research has been initiated by UK-based NGOs (Chilton et al., 2011).

This highlights the possibility of productively framing climate change in terms of other issues which connect with self-transcendence values and intrinsic goals—for example, concern about public health (Nisbet et al., this volume). But it also suggests the imperative to ask how self-transcendence values and intrinsic goals are strengthened more widely in society.

If environmental problems are to be adequately addressed, this will be because self-transcendence values and intrinsic goals come to be held as more important at a cultural level. But this will not be achieved by improving the effectiveness of communications on environmental problems (important as this is). If problems such as climate are to be adequately tackled, public support for the necessary interventions will not be built upon communications about climate change—however sophisticated these may be. It will be built upon the success of strengthening self-transcendence values and intrinsic goals in many areas of citizens' experience.

Communications about climate change—however these are framed—are unlikely, at least today, to have a particularly prominent impact in activating and strengthening, at a cultural level, the values that are embedded in these communications. Climate-related communications are simply not, currently, particularly salient. But citizens' wider understanding of a public health service may be very important in activating the self-transcendence values or intrinsic goals upon which widespread commitment to ambitious action on climate change must come to be premised. So it is one thing for environmental communicators to begin to frame climate change as a threat to public health. As Nisbet et al. argue, such approaches may be productive (especially if they successfully activate concerns relating to the wider community and avoid activating concerns about personal health or security). But environmental organizations must also begin to find common cause with groups that are working to strengthen those public policies (on health or welfare, for example) that are likely to be of the greatest influence in building the wider cultural importance of self-transcendence values and intrinsic goals.

Conversely, because, as we have seen, the importance that an individual places on self-transcendence values tends to be suppressed following activation of self-enhancement values, the envi-

ronment movement must also begin to ask how best to challenge
those influences which currently serve to strengthen self-
enhancement values in society. There is good, though not conclusive, evidence that advertising presents one such influence. It
seems that increased exposure to commercial advertising increases
the importance that individuals place on self-enhancement values,
and that this, in turn, is associated with diminished environmental concern (see Alexander et al., 2011, and references therein). All
advertising, it seems, has the potential to exert important environmental effects arising through the cultural values that the advertising serves to strengthen, irrespective of the direct
environmental impacts of the product that is being promoted. Indeed, as Sinclair and Miller suggest in this volume, "the content of
marketplace advocacy messages poses a significant challenge, with
messages that directly conflict with the argument that major lifestyle and societal changes are required to address current environmental problems" (p. 124).

So Ahern (this volume) is surely right to argue for the importance of considering the ethical issues raised by advertisements
"in terms of the values they directly or indirectly evoke" (p. 184).
He argues that "[t]hese values, not the images, narratives or metaphors used to evoke them, are the subject of the advertising."
Even where these advertisements have not been designed to invoke particular values, it seems that repeated exposure simply to
images of consumer products (estimates suggest that people in the
UK are on average exposed to some 1600 advertisements each day)
may increase the salience of self-enhancement values and suppress pro-social (and, one would predict, pro-environmental) behavior (Bauer et al., 2012). This surely raises ethical questions—
even in the case of purely informational advertisements—in those
instances where it is impossible to remove oneself from exposure to
such advertisements (for example, in the case of advertising in
public spaces) (Alexander et al., 2011). Pressing questions remain
about whether, as Ahern suggests, an advertisement that is "completely conscious/informational," or "that focuses on completely
non-public issues" can be considered entirely ethical. Entirely in-

formational advertisements, focusing on non-public issues, may nonetheless have an important impact in activating particular cultural values.

To call for NGOs to work together to bring particular cultural values to the fore in society is bold. Doubtless, there are extensive opportunities for NGOs to collaborate in such an undertaking, often regardless of differences in their particular issues-based concerns. But nonetheless, this is an ambitious strategy. Given the urgency of addressing challenges like climate change, is it a strategy upon which NGOs should seriously engage? Can it deliver the changes that are needed *quickly* enough?

Certainly, responses to environmental problems must be found *urgently*. But urgency is only part of the story. Such responses must also be:

- *ambitious*—responses must contribute to very significant reductions in environmental impacts. As David MacKay, Chief Scientific Advisor at the UK Department of Energy and Climate Change, has written: "Don't be distracted by the myth that 'every little bit helps.' If everyone does a little, we'll achieve only a little." (MacKay, 2008, p. 114).
- *durable*—it is not enough that new momentum for tackling environmental problems is found soon—this momentum must also be maintained in the long term. Individuals must sustain their private-sphere commitments and persist in exerting pressure on decision-makers—regardless of demographic or economic upheaval. As we've seen, at present, public concern about climate change, for example, is fragile—and has eroded rapidly during the economic recession (Scruggs and Benegal, 2012).
- *systemic*—piecemeal, behavior-by-behavior approaches are inadequate. There is a need for heightened public demand for systemic responses to environmental challenges. In particular, focus must shift from private-sphere behavioral change to various forms of civic engagement—increased motivation to vote differently, lobby political leaders or join demonstrations.

A different approach to communicating and campaigning is needed if urgent, ambitious, durable and systemic change is to be supported. It seems that such an approach will need to be premised upon an appeal to self-transcendent and intrinsic values and strategies designed to bring these values to the fore societally.

My hope is that this volume of essays will contribute to widening awareness of the need for a different approach—and will encourage more researchers to help in its development.

References

Alexander, J., Crompton, T., & Shrubsole, G. (2011). *Think of me as evil? Opening the ethical debates in advertising.* Godalming, Surrey: WWF-UK (Available at www.valuesandframes.org/downloads).

Bauer, M.A., Wilkie, J.E., Kim, J.K., & Bodenhausen, G.V. (2012). Cuing consumerism: Situational materialism undermines personal and social well-being. *Psychological Science.* (Published online 16 March 2012).

Chilton, P., Crompton, T., Kasser, T., Maio, G., & Nolan, A. (2011). *Communicating bigger-than-self problems to extrinsically-oriented audiences.* Godalming, Surrey: WWF-UK (Available at www.valuesandframes.org/downloads).

Crompton, T. (2008). *Weathercocks and signposts: The environment movement at a crossroads.* Godalming, Surrey: WWF-UK (Available at www.valuesand frames.org/downloads).

Crompton, T. (2010). *Common cause: The case for working with our cultural values.* Godalming, Surrey: WWF-UK (Available at www.valuesand frames.org/downloads).

Crompton, T & Kasser, T. (2009). *Meeting environmental challenges: The role of human identity.* Godalming, Surrey: WWF-UK (Available at www.valuesand frames.org/downloads).

Edelman Trust (2011). Edelman Trust Barometer Findings. (Available at: http://www.edelman.com/trust/2011/uploads/edelman%20trust%20barometer %20global%20deck.pdf).

Evans, L., Maio, G.R., Corner, A., Hodgetts, C.J., Hahn, U., & Ahmed, S. (2012). Saving Money While Saving the Environment? Self-Interest and Pro-Environmental Behaviour. Manuscript under review.

Grouzet, F.M.E., Kasser, T., Ahuvia, A., Fernandez-Dols, J.M., Kim, Y., Lau, S., Ryan, R.M., Saunders, S., Schmuck, P., & Sheldon, K.M. (2005). The structure of goal contents across 15 cultures. *Journal of Personality and Social Psychology, 89*, 800–816.

Heath, R.L., Palenchar, M.J., Proutheau, S., & Hocke, T. (2007). Nature, crisis, risk, science, and society: What is our ethical responsibility? *Environmental Communication: A Journal of Nature and Culture, 1*(1), 34–48.

Holmes, T., Blackmore, E., Hawkins, R., & Wakeford, T. (2011). *The Common Cause Handbook,* PIRC: Machynlleth, Wales. (Available at www.valuesand frames.org/downloads).

Hounsham, S. (2006). Painting the town green: How to persuade people to be environmentally friendly. Green Engage. Available at http://www.green-engage.co.uk/PaintingtheTownGreen.pdf.

Kahan, D.M. (2010). Fixing the communications failure. *Nature, 463*, 296–297.

MacKay, D.J.C. (2008). *Sustainable energy—Without the hot air,* Cambridge, UK: UIT.

Schwartz, S.H. (1992). Universals in the content and structure of values: Theory and empirical tests in 20 countries. In M.P. Zanna (Ed.) *Advances in Experimental Social Psychology, 25,* 1–65. New York: Academic Press.

Schwartz, S.H. (2006). Basic human values: Theory, measurement, and applications. *Revue Française de Sociologie, 47*(4), 249–288.

Scruggs, L. & Benegal, S. (2012). Declining public concern about climate change: Can we blame the great recession? *Global Environmental Change, 22,* 505–515.

Sheldon, K.M., Nichols, C.P., & Kasser T. (2011). *Ecopsychology, 3,* 97-104.

Thøgersen, J. & Crompton, T. (2009). Simple and painless? The limitations of spillover in environmental campaigning, *Journal of Consumer Policy, 32,* 141–163.

Vidal, J. (2012). Pollution row after minister deems air quality too costly. *The Guardian,* 27 February. (Available at: http://www.guardian.co.uk/ environment/2012/feb/27/pollution-caroline-spelman)

CONTRIBUTORS

Lee Ahern is assistant professor in the Department of Advertising and Public Relations at Penn State University. His current research focuses on the description, analysis and ethics of strategic messages, primarily in the context of environmental and health communications. Ahern has authored or co-authored more than 25 refereed journal articles, book chapters and conference papers, and is a Senior Research Fellow with the Arthur W. Page Center. His blog posts on environmental communication appear periodically on the Page Center website, CommPro.biz and GreenBiz.com.

Denise Sevick Bortree is an assistant professor in the Department of Advertising and Public Relations at Penn State University. Her research focuses on ethical dimensions of communication in a number of contexts including nonprofit organizations, environmental responsibility, and new media. She has authored more than twenty peer-reviewed journal articles published in journals such as *Journalism and Mass Communication Quarterly, Journal of Public Relations Research, Nonprofit Management and Leadership*, and the *International Journal of Nonprofit and Voluntary Sector Marketing*, among others. Bortree is a Page Legacy Scholar and Senior Research Fellow with the Arthur W. Page Center, and she was a contributor to the United Nations Volunteers' State of the World's Volunteerism report.

Tom Crompton is Change Strategist at WWF-UK, where he leads a program of work examining the importance of cultural values in underpinning public acceptance of—and active demand for—more ambitious action on social and environmental challenges. This programme, called Common Cause, draws on the involvement of a large and rapidly growing number of NGOs and academics. For more information, see www.valuesandframes.org.

Laleah Fernandez is a doctoral candidate in Media and Information Studies at Michigan State University. Fernandez has worked as Assistant to the Editor for *Communication Yearbook*, and as a research assistant for a multi-year project on Communication and Public Will. She recently published on the topic of new media and public will mobilization in the *Journal of Borderland Studies*. Prior to commencing her doctoral studies, Laleah worked in legislative news writing, public opinion polling and political public relations for various public- and private-sector agencies in Michigan.

Harsha Gangadharbatla is an assistant professor in the School of Journalism and Communication at the University of Oregon. His research focuses on new and emerging media, social and economic effects of advertising, and environmental communication. He has authored over 30 publications (journal articles, book chapters, and conference proceedings).

John E. Kotcher is a communications officer in the Office of Communications at the National Academies. In his role at the Academies, John helps to communicate the work of the Academies to broader audiences through public events, and print and online resources. He also works on programs to bring research on science communication to relevant professional communities and develops communication training programs for scientists and engineers. Before joining the National Academies, John attended American University where he obtained a Master of Science in environmental science. John conducted research on climate change communication, examining the potential of opinion leaders to catalyze wider public engagement with the issue and complement traditional media outreach strategies.

Ezra M. Markowitz is currently a post-doctoral research associate at Princeton University. His research centers around the intersection of social and moral psychology, environmental conservation, communications and policy. Current projects include an examination of the "compassion collapse" phenomenon in the environmental domain, an analysis of cross-national climate change threat perceptions, and research on the dynamics and mechanisms of intergenerational environmental stewardship and reciprocity. Markowitz is a National Science Foundation Graduate Research Fellow, a Scholar-in-Residence at American University and a staff member at PolicyInteractive.

Barbara M. Miller is an associate professor in the School of Communications at Elon University. She has worked in public relations/advertising in Virginia and West Virginia and has published research in the *Journal of Advertising, Journal of Public Relations Research, Journal of Applied Communication Research, Journalism and Mass Communication Quarterly, Communication Research Reports, Newspaper Research Journal,* and *Risk Analysis.* She is also the author of *Generating Public Support for Business and Industry: Marketplace Advocacy Campaigns.*

Bernardo H. Motta is an Assistant Professor of Communication Studies at Bridgewater College, Bridgewater, VA. A former lawyer and journalist, his research focuses on issues of environmental risk communication, environmental justice and international environmental law and police issues.

Matthew C. Nisbet is an associate professor of communication and affiliate associate professor of environmental science at American University. Nisbet studies the role of communication and media in policymaking and public affairs, focusing on science-related controversies. In this area, he is the author of more than 50 peer-reviewed studies, scholarly book chapters, and monographs. A 2011 editorial at the journal *Nature* cited his research as "essential reading for anyone with a passing interest in the climate change debate." Nisbet has served as a health policy investigator with the Robert Wood Johnson Foundation, a Google Science Communication Fellow, and a Shorenstein Fellow in Press, Politics, and Public Policy at Harvard University.

Michael J. Palenchar is an associate professor and co-director of the Risk, Health and Crisis Communication Research Unit at the University of Tennessee. A former public relations practitioner, his research focuses on matters of risk and crisis communication, issues management and public relations.

Charles T. Salmon is Professor of Communication Studies at the Wee Kim Wee School of Communication and Information, Nanyang Technological University, Singapore. Dr. Salmon's research focuses on issues at the intersection of public information, public health, and public opinion, including environmental communication. He has taught courses on public relations, public opinion, social marketing and health communication.

Kim Sheehan is professor of advertising at the University of Oregon, and also serves as Director of the Master's Program in Strategic Communication. She brings over 12 years of experience in advertising and marketing to the advertising sequence. Her research involves culture and new technology, and she has published extensively about social media, online privacy, green advertising, advertising ethics, and Direct-to-Consumer prescription drug advertising. She is the author or co-author of six books, and her research has appeared in the *Journal of Advertising* and the *Journal of Public Policy and Marketing*, among others.

Brant Short is Professor of Communication Studies at Northern Arizona University. He is author of *Ronald Reagan and the Public Lands: America's Conservation Debate*, and has published scholarship on popular culture, environmental politics, and rhetoric. His work has appeared in *Communication Studies*, the *Southern Communication Journal*, the *Western Communication Journal, Presidential Studies Quarterly*, the *Journal of Religion and Popular Culture*, the *Environmental Green Journal*, and the *Journal of Communication and Religion*, among others.

Janas Sinclair is an associate professor in the School of Journalism and Mass Communication at the University of North Carolina at Chapel Hill. She teaches courses on advertising and marketing communication. Her research focuses on psychological processing of persuasive messages and the antecedents of trust, particularly for messages regarding the environment and technology. Her research has been published in advertising, mass communication, and science communication journals including the *Journal of Advertising, Risk Analysis, Science Communication,* and *Journalism and Mass Communication Quarterly.*

INDEX